BIOTECHNOLOGY

BIOTECHNOLOGY

M. Selfi

RAJAT PUBLICATIONS
New Delhi

Biotechnology
by *M. Selfi*

This edition is published by Rajat Publications
4740/23, Ansari Road Daryaganj,
New Delhi - 110002

ISBN: 978-81-7880-250-3

Edition 2026

Printed and bound in India

RAJAT
PUBLICATIONS

4740/23, Ansari Road, Daryaganj,
New Delhi- 110002 (India)
Mob:- +91-9811484973, +91-9811105389
E-mail: rajat_publication@yahoo.com

Preface

This readable text provides the essentials of biotechnology at a level that is neither two high (for novice students) nor too for (for advances students). This has been done for the first time and substantially distinguishes the present textbook from those issued previously on the subject. The concepts included in this text acquaint students with modern concepts of biotechnology, a general of which is needed for their future work. In attempting this text, the author has been guided by an innate desire to bring home the various modern concepts of biotechnology in as lucid and logical a manner as possible. A precise knowledge of one's own biological and chemistry background is indispensable for comprehending the concepts included in this text.

—**Editor**

Contents

1

Introduction to Biotechnology

Introduction

Biotechnology is *technology* based on *biology*, especially when used in *agriculture*, *food science*, and *medicine*.

Of the many different definitions available the one formulated by the UN "Convention on *Biological Diversity*" is the most all-encompassing:

> "Biotechnology is any technological application that uses biological systems, living organisms, or derivatives thereof, to make or modify products or processes for specific use."

One section of biotechnology is the directed use of *organisms* by *humans* for production (beer, *milk*-products, skin). Naturally present *bacteria* are also involved in the mining industry in *bioleaching*. Other uses of biotechnology involve recycling, treatment of waste, cleaning up sites contaminated by industrial activities (*bioremediation*) or production of *biowar* agents.

There are also applications of biotechnology that do not use living *organisms*. An example are *DNA chips* used in genetics, or radioactive tracers used in medicine.

Although biotechnology is publicly associated with *cloning* and *genetic engineering*, the goal of biotechnology is to advance the tools of medicine and solve problems related to the production of biologically derived products, not the whimsical manipulation of life.

Today, biotechnology, or modern biotechnology, is often associated to the use of genetically altered *microorganisms* such as E. coli or yeast for producing substances like *insulin* or *antibiotics*. It can also refer to *transgenic animals* or *transgenic plants*, such as Bt corn. Genetically altered Mammalian cells, such as *Chinese Hamster* ovarian cells, are also widely used to manufacture pharmaceuticals.

Sub-Fields of Biotechnology

There are number of jargon terms for sub-fields of biotechnology.

Red biotechnology is biotechnology applied to *medical* processes. An example would include an organism designed to produce an *antibiotic*, or engineering genetic cures to diseases through *genomic manipulation*.

White biotechnology, also known as grey biotechnology, is biotechnology applied to *industrial* processes. An example would include an organism designed to produce a useful chemical.

White biotechnology tends to consume less resources that traditional processes when used to produce industrial goods.

Green biotechnology is biotechnology applied to *agricultural* processes. An example would include an organism designed to grow under specific environmental conditions or in the presence (or absence) of certain agricultural chemicals.

Green biotechnology tends to produce more environmentally friendly solutions then traditional industrial agriculture. An example of this would include a plant engineered to express a *pesticide*, thereby eliminating the need for external application of pesticides.

The term blue biotechnology has also been used to describe the marine and aquatic applications of biotechnology, but its use is relatively rare.

Biotechnology Timeline

- 8000BC collecting of seeds for replanting. Evidence that Babylonians, Egyptians and Romans used selective breeding practices to improve *livestock*. (*Artificial selection*)

- 6000BC brewing *beer*, fermenting *wine*, baking *bread* with help of yeast
- 4000BC Chinese made *yoghurt* and cheese with *lactic-acid-producing bacteria*
- 1500 plant collecting around the world
- 1300 *Nikolai I. Vavilov* created comprehensive research on breeding animals
- 1880 microorganisms discovered
- 1856 *Gregor Mendel* started recombinant plant genetics
- 1919 *Karl Ereky*, a Hungarian engineer, first used the word biotechnology
- 1980 modern biotech is characterized by *recombinant DNA technology*. The *prokaryote* model, *E. coil*, is used to produce *insulin* and other medicine, in human form. (About 5% of diabetics are allergic to animal insulins available before)
- 1992 FDA approves of the first GM food from *Calgene*: "flavor saver" tomato
- 2000 completion of the *Human Genome Project*

2

Concepts in Biotechnology

Introduction

It is any technique that is used to make or modify the products of living organisms in order to improve plants or animals or to develop useful microorganisms. By this definition, biotechnology has actually been practised for centuries, as exemplified by the use of yeast and bacteria in the production of various foods, such as wine, bread, and cheese. However, in modern terms, biotechnology has come to mean the use of cell and tissue culture, cell fusion, molecular biology, and in particular, recombinant deoxyribonucleic acid (DNA) technology to generate unique organisms with new traits or organisms that have the potential to produce specific products. Some examples of products in a number of important disciplines are described below.

Genetics

Recombinant DNA technology opened new horizons in the study of gene function and the regulation of gene action. In particular, the ability to insert genes and their controlling nucleic acid sequences into new recipient organisms allows for the manipulation of these genes in order to examine their activity in unique environments, away from the constraints posed in their normal host. Transformed plants, animals, yeast, and bacterial genes may be examined in this way.

Microbiology

Genetic transformation normally is achieved easily with microorganisms; new genetic material may be inserted into them, either into their chromosomes or into extrachromosomal elements, the plasmids. Thus, bacteria and yeast can be created to metabolize specific products or to produce new products. Concomitant technologies have been developed to scale up the production of the microorganisms to generate products, such as enzymes, carbohydrates, and proteins, in great quantity.

Immunology

Genetic engineering has allowed for significant advances in the understanding of the structure and mode of action of antibody molecules. Practical use of immunological techniques is pervasive in biotechnology. Notably, antibodies are used in diagnostic procedures for detecting diseases of plants and animals, and in detecting minute amounts of such materials as toxic wastes, drugs, and pesticides. The antibodies themselves are employed to target therapeutic agents to specific cellular sites. Antibodies are bivalent molecules that bind to their target molecules at one or both of their two combining sites. Hybrid antibodies have been produced in which one of the sites contains a drug or a poison while the other site directs the antibody to its target, that is, a cancerous cell. The ability to artificially combine subunits of antibodies produced in different species also will tailor them for specific targets. Antibodies can be prepared that have enzymatic properties, thereby enabling them to deactivate target molecules with which they combine in a cell.

Monoclonal antibodies respond to a single antigenic site, allowing for great specificity. They have been most important in the diagnostic arena, where tests have been developed for human, plant, and animal diseases and for pregnancy and ovulation prediction.

Agriculture

Agriculture is concerned with producing plants that are tolerant to specific herbicides. This tolerance would allow crops to be sprayed with the particular herbicide, and only the weeds would be killed, not the genetically engineered crop species. Some herbicide-tolerant crop species have been isolated by selection of tissue culture variants, but in other cases plants have been transformed with genes from bacteria that detoxify the herbicide. For example, a gene isolated from the soil bacterium *Klebsiella ozaenae* has been used to create plants that are resistant to the herbicide bromoxynil. Other strategies have been to alter plants so that they will overproduce the herbicide-sensitive biochemical target, or so that the biochemical target will be altered, thereby reducing the affinity of the herbicide for its biochemical target in the crop species.

Tolerance to plant virus diseases has been induced in a number of crop species by transforming plants with portions of the viral genome, in particular the virus's coat protein and the replicase enzyme used for virus multiplication. Plants exhibiting this tolerance are available for commercial use once regulatory constraints are removed. Genes coding for protein toxins from the bacterium *Bacillus thuringiensis* have been introduced into a number of plant species to provide insect tolerance. Potato plants tolerant to the Colorado potato beetle and cotton plants tolerant to the cotton boll worm were the first commercial products of this technology.

There are a number of techniques that can be used to enhance the quality and nutrition of foods. Recombinant DNA strategies are used to retard the softening of tomatoes so they can reach the consumer with better flavor and keeping qualities. There is a technique to increase the starch content of the potato to enhance the quality of french fries and potato chips, including reduction of their capacity to absorb oils during processing. With both selective plant breeding and genetic engineering, oil seed rape (also known as canola or *Brassica napus)* has been produced to yield plant-based oils (oleochemicals) for the specialty chemicals industry. Such oils are used to produce amides, fatty acids, esters, and other chemical intermediates. Oils high in erucic acid are used as lubricants and as additives to automatic transmission fluid. Oil seed rape is also an important source of oil for margarine production and cooking. Genetic engineering techniques can also be used to lower the proportion of saturated fat by inserting the gene for the stearoyl-acyl carrier protein desaturase enzyme gene into oil seed rape and other oil-producing crop plants.

Biotechnology also holds great promise in the production of vaccines for use in maintaining the health of animals. Subunit vaccines could replace live-virus or killed-whole-virus vaccines. For example, feline leukemia is estimated to infect 1.5 million cats annually in the United States. Conventional vaccines are ineffective against the virus and may actually spread the disease, but the subunit vaccines that have been developed overcome these disadvantages. Another example is pseudorabies virus, the causal agent of a disease affecting hogs. The conventional killed-virus vaccine gives only partial protection, and a live-virus vaccine is only slightly more effective, but there is a genetically engineered vaccine. This vaccine reduces virus virulence because the gene producing the enzyme that enables the virus to escape from nervous tissue has been deleted. Another vaccine incorporates genes for pseudorabies proteins into viable vaccinia virus as a vector to immunize animals. Recombinant vaccines have also been produced for foot-and-mouth virus and parvovirus.

Interferons may be useful in treating bovine shipping fever, a complex of respiratory infections manifested during the crowded conditions experienced by cattle when shipped. Recombinant-produced bovine growth hormone is effective in increasing the production of milk in cows and in reducing the ratio of fat to lean meat in beef cattle.

Animals may be transformed to carry genes from other species, including humans, and are being used to produce valuable drugs. This technology has been termed biopharming. For example, goats are used to produce tissue plasminogen activator, which has been effective in dissolving blood clots. Transgenic animals have been produced that carry the acquired immune deficiency syndrome (AIDS) virus, providing a practical cost-effective experimental model system to test control measures for the disease.

Plant Science

Plants can be transformed quite easily to enable them to express foreign genes. This field developed very rapidly after the first transformation of a plant was reported in 1982, and a number of transformation procedures are now available. The plasmid derived from the plant pathogenic bacterium *Agrobacterium tumefaciens* is widely used as a transformation vector. Another method involves shooting DNA-coated tungsten (or gold) particles into cells. Controlling genetic elements have been identified that allow for the insertion of genes into specific tissues or plant organs. For example, nucleic acid sequences may be targeted to the pollen grain to specifically inactivate genes involved in producing pollen, thereby allowing for the production of sterile male plants. Such a technique can be useful in the commercial production of hybrid seeds of crop species.

Coupled with these transformation procedures has been the development of tissue culture techniques to enable the transformed cells to be regenerated into whole plants. Many plant species have been regenerated by this technique, thereby facilitating the transfer of useful genes to the most important crops (rice, maize, wheat, and potatoes). Cultures of plant cells, roots, and tissues are used to produce secondary plant products, such as the anticancer drug taxol.

Medicine

Genetic engineering has enabled the large-scale production of proteins which have great potential for treatment of heart attacks. One of these drugs is tissue plasminogen activator, which is effective in dissolving blood dots. Active tissue plasminogen activator, a very minor constituent of human blood vessels and some other tissues, can be produced by recombinant technology in transformed tissue culture cells or in the filamentous fungus *Aspergillus nidulans*. Urokinase is produced by recombinant technology; it has potential for the treatment of heart attacks is urokinase.

Human gene products, produced with genetic engineering technology, may have potential use as commercial drugs. Cloned human growth hormone is being used for the treatment of childhood dwarfism. Epidermal growth factor, a protein that causes the replication of epidermal cells, has applications to wound healing.

Recombinant technology has also been employed to produce vaccines from subunits of viruses, so that the use of either live or inactivated viruses as immunizing agents is avoided. Conventional vaccines are sometimes infectious in themselves, and most require refrigeration which makes their use in tropical countries a problem.

Poliovirus vaccines involving attenuated live virus are a source of the perpetuation of the disease due to rare mutations of the virus to virulence. Recombinant technology is being used to modify the viral genome to prevent reversion. Malaria, one of the most important parasitic diseases of humans, is another case in which biotechnology is being applied to produce a vaccine. Proteins from the causal organism, *Plasmodium sp.*, are poorly immunogenic. To enhance immunogenicity, they are transferred to the attenuated bacterium *Salmonella typhimurium*, which is then used as an immunizing agent. Another strategy has been to fuse the gene for the immunological determinant to that of a hepatitis B virus protein, to express the fusion protein in yeast, and to use that chimeric protein as a vaccine.

Cloned genes and specific, defined nucleic acid sequences can be used as a means of diagnosing infectious diseases or in identifying individuals with the potential for genetic disease. The specific nucleic acids used as probes are normally tagged with radioisotopes, and the DNAs of candidate individuals are tested by hybridization to the labeled probe. The technique has been used to detect latent viruses such as herpes, bacteria, mycoplasmas, and plasmodia, and to identify Huntington's disease, cystic fibrosis, and Duchenne muscular dystrophy. In some cases, restriction-length polymorphisms are useful. When DNA is cut into small fragments by specific restriction enzymes and then is probed with specific genes or nucleic acids, differences between individuals in a population can be identified, and the relationships of specific patterns to specific diseases or traits can be determined. This technology is also used for identifying important genes in plants and animals in order to improve breeding stocks. The technique known as DNA fingerprinting is another application of the technology, specifically as a forensic tool. It allows a specific identity to be assigned to individuals through their DNA in much the same way that fingerprints are used. The technique has the potential of distinguishing the DNA from one individual in a population of 10 billion. Tissue samples containing DNA left at the scene of a crime, such as bone, blood, semen, skin, hair (if it is attached to its root), saliva, and sweat, can be used in the procedure. If the amount of DNA found is too small to be useful, the technique of polymerase chain reaction can amplify it to practical levels.

Gene functions can often be blocked by attacking them with complementary or antisense sequences of the same gene. This is valuable in defining functions for specific genes, but it also has a number of practical applications. In agriculture it has been used to generate male sterile plants, enabling the production of hybrid varieties more easily, and to slow the ripening of tomatoes. Antisense technology also presents the opportunity for useful gene therapy. For example, the human immunodeficiency virus (HIV) can be inhibited by transforming T lymphocytes with antisense nucleic acids directed against a virus enzyme, reverse transcriptase.

It is also possible to put foreign genes into cells and to target them to specific regions of the recipient genome. This presents the possibility of developing specific therapies for hereditary diseases, exemplified by sickle-cell anemia, which is caused by a defect in the β-globin gene which results in defective hemoglobin in affected individuals.

Environment

Microorganisms, either genetically engineered or selected from natural populations, are used to degrade toxic wastes in the environment. For example, polycyclic aromatic compounds, such as polychlorinated biphenyls, and petroleum products which contaminate soil and ground-water supplies may be degraded by populations of microorganisms. These technologies have the potential to solve some significant environmental problems. Waste products of industry and agriculture are being composted, with added microorganisms selected for their capacity to degrade organic materials.

3

Genetic Engineering

Introduction

It is the artificial recombination of nucleic acid molecules in the test tube, their insertion into a virus, bacterial plasmid, or other vector system, and the subsequent incorporation of the chimeric molecules into a host organism in which they are capable of continued propagation. The construction of such molecules has also been termed gene manipulation because it usually involves the production of novel genetic combinations by biochemical means.

Genetic engineering provides the ability to propagate and grow in bulk a line of genetically identical organisms, all containing the same artificially recombinant molecule. Any genetic segment as well as the gene product encoded by it can therefore potentially be amplified. For these reasons the process has also been termed molecular cloning or gene cloning.

Basic Techniques

The central techniques of such gene manipulation involve (1) the isolation of a specific deoxyribonucleic acid (DNA) molecule or molecules to be replicated (the passenger DNA); (2) the joining of this DNA with a DNA vector (also known as a vehicle or a replicon) capable of autonomous replication in a living cell after foreign DNA has been inserted into it; and (3) the transfer, via transformation or transfection, of the recombinant molecule into a suitable host.

Isolation of Passenger DNA

Passenger DNA may be isolated in a number of ways; the most common of these involves DNA restriction.

DNA Restriction

DNA purified from an organism can be prepared for cloning only after it has been cut into smaller molecules. Restriction endonucleases make possible the cleavage of high-molecular-weight DNA. Although three different classes of these enzymes have been described, only type II restriction endonucleases have been used extensively in the manipulation of DNA. Type II restriction endonucleases are DNAases that recognize specific short nucleotide sequences (usually 4 to 6 base pairs in length), and then cleave both strands of the DNA duplex, generating discrete DNA fragments of defined length and sequence.

A number of restriction enzymes make staggered cuts in the two DNA strands, generating singlestranded termini. Restriction enzyme Eco R1 cleaves the sequence GAATTC (where G = guanine; A = adenine; T = thymine; C = cytosine) between the G and A residues, generating protruding single-stranded ends having 5′ termini.

The various fragments generated when a specific DNA is cut by a restriction enzyme can be easily resolved as bands of distinct molecular weights by agarose gel electrophoresis. Specific sequences of these bands can be identified by a technique known as Southern blotting. In this technique, DNA restriction fragments resolved on a gel are denatured and blotted onto a nitrocellulose filter. The filter is incubated together with a radioactively labeled DNA or RNA probe specific for the gene under study. The labeled probe hybridizes to its complement in the restricted DNA, and the regions of hybridization are detected autoradiographically. Fragments of interest can then be eluted out of these gels and used for cloning. Purification of particular DNA segments prior to cloning reduces the number of recombinants that must later be screened.

Mechanical Shearing of DNA

Another method that has been used to generate small DNA

fragments is mechanical shearing. Intense sonification of high-molecular-weight DNA with ultrasound, or high-speed stirring in a blender can both be used to produce DNA fragments of a certain size range. Shearing results in random breakage of DNA, producing termini consisting of short, single-stranded regions.

Other Sources

Another source of DNA to be replicated is DNA complementary to poly(A) RNA, or cDNA, which is synthesized in the test tube. The cDNA synthesis involves the use of the enzyme, reverse transcriptase, which uses poly(A) RNA as a template for the synthesis of a complementary DNA strand. The RNA strand in the resulting RNA-DNA hybrid is removed by alkaline hydrolysis, and the remaining singlestranded DNA is converted into the double-stranded form in a second reaction involving DNA polymerase I.

In addition, short oligonucleotides, synthesized chemically, can be used for cloning purposes.

Joining DNA Molecules

Once the proper DNA fragments have been obtained, they must be joined. This is accomplished by the following methods.

Cohesive End Ligation

When cleavage with a restriction endonuclease creates cohesive ends, these can be annealed with a similarly cleaved DNA from another source, including a vector molecule. When such molecules associate, the joint has nicks a few base pairs apart in opposite strands. The enzyme DNA ligase can then repair these nicks to form an intact, duplex recombinant molecule, which can be used for transformation and the subsequent selection of cells containing the recombinant molecule (Fig. 3.1).

Cohesive ends can also be created by the addition of synthetic DNA linkers to blunt-ended DNA molecules. These linkers are short DNA duplexes containing the recognition site for a restriction enzyme which produces cohesive termini. Linkers are ligated to blunt-ended passenger DNA molecules by DNA ligase encoded

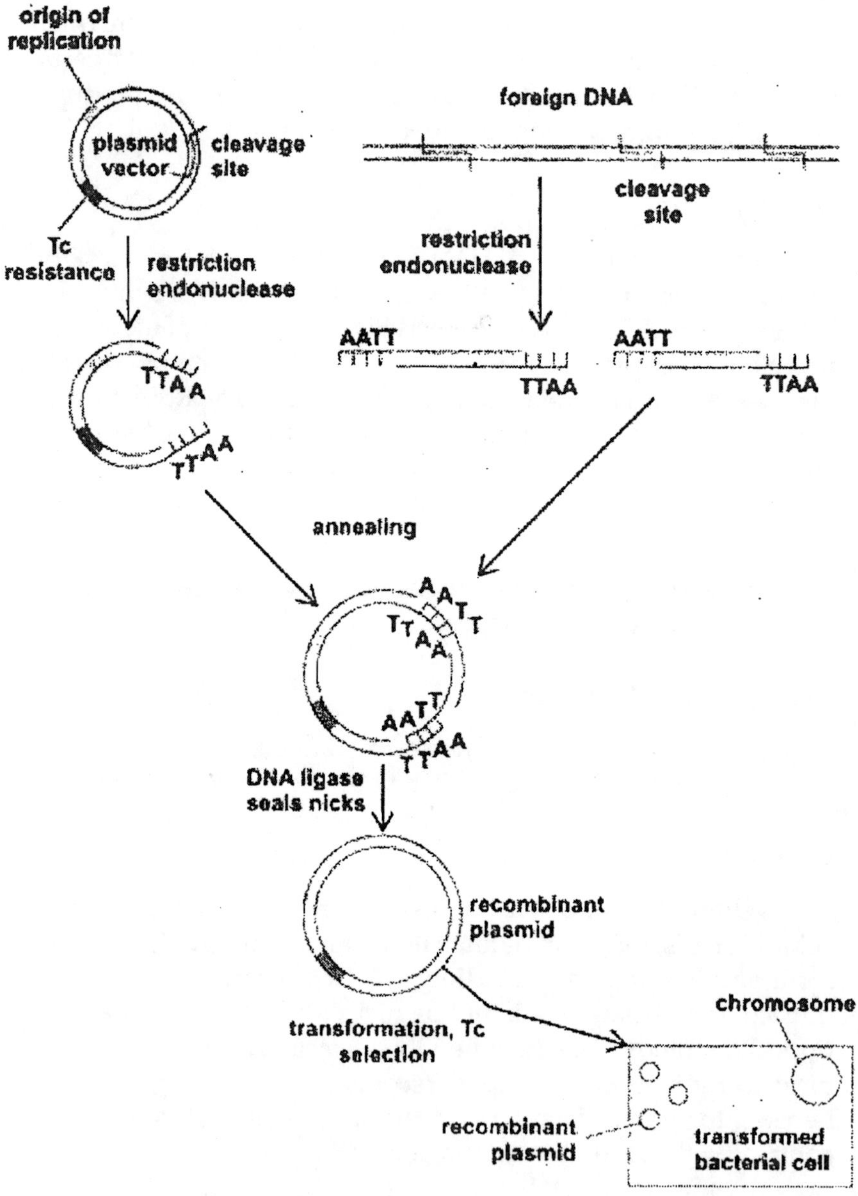

Fig. 3.1. Use of DNA ligase to create a covalent DNA recombinant joined through association of termini generated by Eco R1. (*After R. W. Old and S. B. Primrose, Principles of Gene Manipulation, 3d ed., Blackwell Scientific Publications, 1985*)

by the phage T4. After digesting the product with the restriction enzyme that cleaves the linkers, the products can be ligated to any vehicle DNA via the complementary termini and then cloned.

Homopolymer Tailing

Another method for joining DNA molecules involves the addition of homopolymer extensions to different DNA populations followed by an annealing of complementary homopolymer sequences. For example, short nucleotide sequences of pure adenine can be added to the 3′ ends of one population of DNA molecules and short thymine blocks to the 3′ ends of another population. The two types of molecules can then anneal to form mixed dimeric circles that can be used directly for transformation. Single-stranded gaps that may remain in the two strands at each connection will be repaired in the transformed cell.

Blunt-End Ligation

T4 DNA ligase carries out the intermolecular joining of DNA substrates at completely base-paired ends; such blunt ends can be produced by cleavage with a restriction enzyme or by mechanical shearing followed by enzyme treatment.

Transformation

The desired DNA sequence, once attached to a DNA vector, must be transferred to a suitable host. Transformation is defined as the introduction of foreign DNA into a recipient cell. Transformation of a cell with DNA from a virus is usually referred to as transfection.

Transformation in any organism involves (1) a method that allows the introduction of DNA into the cell and (2) the stable integration of DNA into a chromosome, or maintenance of the DNA as a self-replicating entity.

Prokaryotes and Lower Eukaryotes

Escherichia coli is usually the host of choice for cloning experiments, and transformation of *E. coli* is an essential step in these experiments. *Escherichia coli* treated with calcium chloride are able to take up DNA from bacteriophage lambda as well as plasmid

DNA. Calcium chloride is thought to effect some structural alterations in the bacterial cell wall.

In addition to *E. coli,* other prokaryotes have also been used as hosts for cloning experiments, among them the nonenteric bacteria, *Bacillus* species, and actinomycetes. An efficient method for transformation in *Bacillus* involves polyethylene glycol-induced DNA uptake in bacterial protoplasts and subsequent regeneration of the bacterial cell wall. Actinomycetes can be similarly transformed. Transformation can also be achieved by first entrapping the DNA with liposomes followed by their fusion with the host cell membrane. Similar transformation methods have been developed for lower eukaryotes such as the yeast *Saccharomyces cerevisiae* and the filamentous fungus *Neurospora crassa.*

Animal Cells

Several methods are available for the transfer of DNA into cells of higher eukaryotes. Specific genes or entire viral genomes can be introduced into cultured mammalian cells in the form of a coprecipitate with calcium phosphate. DNA complexed with calcium phosphate is readily taken up and expressed by mammalian cells. DNA complexed with diethylamino-ethyl-dextran (DEAE-dextran), or DNA trapped in liposomes or erythrocyte ghosts may also be used in mammalian transformation. Alternatively, bacterial protoplasts containing plasmids can be fused to intact animal cells with the aid of chemical agents such as polyethylene glycol (PEG). Finally, DNA can be directly introduced into cells by microinjection. The efficiency of transfer by each of these methods is quite variable.

Higher Plants

Introduction of DNA sequences by insertion into the transforming (T)-DNA region of the tumor-inducing (Ti) plasmid of *Agrobacterium tumefaciens is* a method of introducing DNA into plant cells and ensuring its integration. The Ti plasmid has been used to introduce selectable marker genes, including bacterial genes coding for antibiotic resistance, into plant cells. The control sequences used to express these chimeric genes are derived from plant genes. Many foreign genes have been expressed in the transformant plant cells and have conferred selectable properties into plant cells.

Although the Ti plasmid can be used as a gene transfer vector, its use is limited in that *A. tumefaciens*, the bacterium that transmits Ti plasmids, normally infects primarily dicotyledonous plants. Monocotyledonous plants, including economically important cereals, cannot be infected by the Ti plasmid.

Because of the limitations of the host range of A. *tumefaciens*, alternative transformation systems are being developed for gene transfer in plants. They include the use of liposomes, as well as induction of DNA uptake in plant protoplasts. With those methods, however, transformation frequencies are low. Plant DNA viruses, such as the cauliflower mosaic virus, might also be used as vectors for DNA transfer.

Foreign DNA has been introduced into plant cells by a technique called electroporation. This technique involves the use of electric pulses to make plant plasma membranes permeable to plasmid DNA molecules. Plasmid DNA taken up in this way has been shown to be stably inherited and expressed.

Cloning Vectors

There is a large variety of potential vectors for cloned genes. The vectors differ in different classes of organisms.

Prokaryotes and Lower Eukaryotes

Three types of vectors have been used in these organisms: plasmids, bacteriophages, and cosmids.

Plasmids

Plasmids are extrachromosomal DNA sequences that are stably inherited. *Escherichia coli* and its plasmids constitute the most versatile type of host-vector system known for DNA cloning.

Several natural plasmids, such as ColE1, have been used as cloning vehicles in *E. coli*. In addition, a variety of derivatives of natural plasmids have been constructed by combining DNA segments and desirable qualities of older cloning vehicles. These qualities, which facilitate manipulation of cloned DNA and limit the ability of recombinants to survive in the environment beyond the laboratory, include (1) reduced molecular weight of the plasmid; (2) construction of only one site for a given restriction enzyme located in a region not essential for plasmid replication; (3) introduction of one or more phenotypic markers into the plasmid to permit selection of cells containing the plasmid and to aid in the identification of plasmids that have incorporated the foreign DNA. It is also desirable to use plasmids that can be obtained in high yields.

The most versatile and widely used of these plasmids is pBR322. This plasmid has been completely sequenced, providing a detailed restriction enzyme cleavage map and DNA sequence information (Fig. 3.2). The pBR322 plasmid carries the genes coding for resistance to the antibiotics ampicillin and tetracyline.

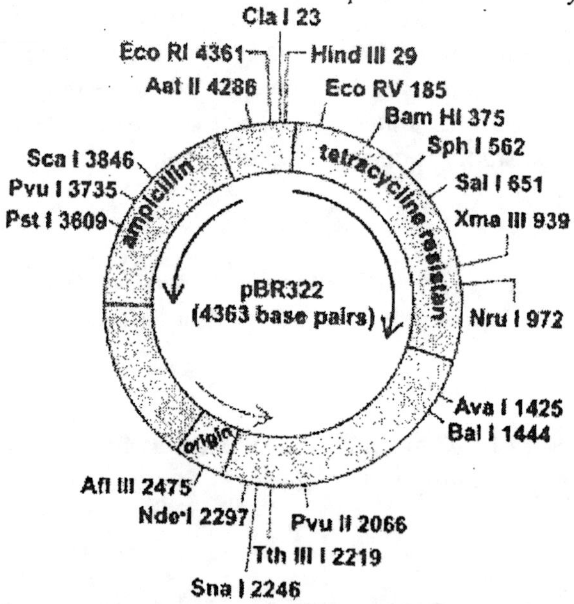

Fig. 3.2. Structure of pBR322 showing the unique cleavage sites. The numbers relate to the coordinate of the base which corresponds to the 5 nucleotide of each recognition sequence. The black arrows show the directions of transcription of the Ap^R and Tc^R genes. The colored arrow shows the direction of DNA replication. (*After R. W. Old and S. B. Primrose, Principles of Gene Manipulation, 3d ed., Blackwell Scientific: Publications, 1985*).

Transformation in yeast has been demonstrated using a number of plasmids, including vectors derived from the naturally occurring 2μ plasmid of yeast. These vectors contain DNA fragments from the 2μ yeast plasmid, yeast nuclear DNA, and the *E. coli* vector, pMB9. Other yeast vectors are recombinants consisting of *E. coli* vectors and yeast DNA fragments.

Bacteriophages

Bacteriophage lambda is a virus of *E. coli*. Several lambda-derived vectors have been developed for cloning in *E. coli*, and for the isolation of particular genes from eukaryotic genomes. These lambda derivatives have several advantages over plasmids: (1) Thousands of recombinant phage plaques can easily be screened for a particular DNA sequence on a single petri dish by molecular hybridization. (2) Packaging of recombinant DNA in laboratory cultures provides a very efficient means of DNA uptake by the bacteria. (3) Thousands of independently packaged recombinant phages can be easily replicated and stored in a single solution as a "library" of genomic sequences.

Cosmids

Plasmids have also been constructed that contain the phage *cos* DNA site, required for packaging into the phage particles, and ColE1 DNA segments, required for plasmid replication. These plasmids have been termed cosmids. After packaging laboratory cultures, the vector is used to infect a host. The recombinant cosmid DNA is injected and circularizes like phage DNA but replicates as a plasmid. Transformed cells are selected on the basis of a vector drug resistance marker. Cosmids provide an efficient means of cloning large pieces of foreign DNA and have therefore been used for constructing libraries of eukaryotic genomic fragments.

Animal Cells

In contrast to the wide variety of plasmid and phage vectors available for cloning in prokaryotic cells, relatively few vectors are available for introducing foreign genes into animal cells. In the latter case, the most commonly used vectors are derived from simian virus 40 (SV40).

The SV40 genome consists of a small covalently closed circular DNA molecule for which the entire nucleotide sequence is known. Normal SV40 cannot be used as a vector, since there is a physical limit to the amount of DNA that can be packaged into the virus capsid, and the addition of foreign DNA would generate a DNA molecule too large to be packaged into a viral particle. However, SV40 mutants lacking portions of the genome can be propagated in mixed infections in which a "helper" virus supplies the missing function.

Propagating recombinants in eukaryotic cells using SV40 vectors imposes some limitations, such as the severe limitation on the size of gene segments that can be cloned in the small viral genome and the fact that only cells permissive for SV40 replication can be used.

Bovine papilloma virus is another DNA virus that has been used as a vector in eukaryotic animal cells. The rat insulin gene inserted into the viral DNA has been used successfully to transform mouse cells.

Another system for the stable transfer of genes into mammalian cells involves cotransformation of cells with two physically separate sets of genes. For example, the introduction of the thymidine kinase (tk) gene from Herpes simplex virus to mouse tk^- cells results in stable and efficient transformation of cells that express the viral gene.

In *Drosophilia*, transposable DNA elements, that is, DNA segments that are capable of moving from one position to another within the genome of a cell, have provided the basis for development of an efficient and controlled system of gene transfer. DNA segments of interest can be transposed into germ-line chromosomes along with a transposable element to which the segment has been previously ligated.

Plant Cells

Two systems for the delivery and integration of foreign genes into the plant genome are the Ti plasmid of the soil bacterium *Agrobacterium* and the DNA plant virion cauliflower mosaic virus.

The Ti plasmid is a natural gene transfer vector carried by *A.*

tumefaciens, a pathogenic bacterium that causes crown gall tumor formation in dicotyledonous plants. A T-DNA segment present in the Ti plasmid becomes stably integrated into the plant cell genome during infection. This property of the Ti plasmid has been exploited to show that DNA segments inserted in the T-DNA region can be cotransferred to plant DNA.

Caulimoviruses belong to a group of plant DNA viruses that contain double-stranded DNA. Of these, cauliflower mosaic virus is the best studied for its cloning vehicle potential. The genomes of over 30 strains of cauliflower mosaic virus have been mapped by using restriction enzymes. Purified preparations of the cauliflower mosaic virus DNA can infect plant cells with relatively high efficiency. Cauliflower mosaic virus DNA is also transformable after gene manipulation, which indicates that it might be possible to use cauliflower mosaic virus as a plant vector.

Cloned Gene Expression

It is sometimes the aim of the genetic engineer to promote the expression of a cloned gene not only for the analysis of gene structure and function but also for amplification of the synthesis of a desirable gene product. DNA cloning methods have enabled the genetic modification of bacteria and unicellular eukaryotes, rendering them capable of producing virtually any gene product of animal or plant cells.

The aim of gene transfer experiments is to examine the mechanisms of gene regulation and expression in the context of normal cell differentiation. DNA sequences cloned by recombinant DNA technology have been transferred into fertilized mouse, *Drosophila*, and *Xenopus* oocytes by microinjection techniques. In the mouse and *Drosophila* this system has allowed study of the expression and regulation of transferred gene sequences in specific tissues during normal cellular differentiation, and their transmission from one generation to the next. The *Xenopus* oocyte has been used mainly for analysis of transcription and translation of injected sequences.

In the mouse, foreign DNA is injected into a pronucleus of a fertilized egg and subsequently becomes incorporated into the chromosomes of the diploid zygote. The injected eggs are then transferred to a foster-mother mouse, where normal embryonic development occurs. Some of the progeny mice come to contain the foreign DNA in their cells, and the expression of the introduced DNA can be examined during development of these transgenic mice.

A variety of recombinant DNAs have been injected into the mouse oocyte, including plasmids containing SV40 sequences, the Herpes simplex tk gene, a human alpha interferon cDNA, and beta-globin genes from a variety of animals. The injected DNA has been shown to be integrated into the genomes of some of the transformants, and in many cases transmission of DNA from the original transgenic mouse to its progeny has also been demonstrated.

Applications

Genetic engineering technology has made possible a wide variety of practical applications, several of which are discussed below.

Isolation of Specific Genes

Recombinant DNA technology has permitted the isolation and detailed structural analysis of a large number of prokaryotic and eukaryotic genes. This contribution is especially significant in the eukaryotes because of their large genomes. The methods outlined above provide a means of fractionating and isolating individual genes, since each clone contains a single sequence or a few DNA sequences from a very large genome. Isolation of a particular sequence of interest has been facilitated by the ability to generate a large number of clones and to screen them with the appropriate "probe" (radioactively labeled RNA or DNA) molecules.

Analysis of Gene Structure and Function

Genetic engineering techniques provide pure DNAs in amounts sufficient for mapping, sequencing, and direct structural analyses. Furthermore, gene structure-function relationships can be studied by reintroducing the cloned gene into a eukaryotic nucleus and assaying for transcriptional and translational activities.

The DNA sequences can be altered by mutagenesis, before their reintroduction in order to define precise functional regions.

Such analyses have provided insights into the complexity of the eukaryotic genome. For example, many eukaryotic genes contain noncoding sequences (introns) interspersed in the coding sequences (exons) of genes. The primary RNA transcripts are spliced to remove the noncoding sequences as the mRNA matures. Exon-intron boundaries from diverse sources have been sequenced to reveal a common sequence pattern at the opposite ends of different introns. These sequences may be part of a mechanism that brings opposite ends of introns together prior to splicing.

Cloning and sequencing of increasing numbers of eukaryotic genes have brought to light similarities in specific sequences near the 5′ and 3′ ends of different genes. These sequences are suspected to be involved in the regulation of transcription or translation.

Molecular probes for a number of eukaryotic genes have revealed the existence of clustered gene families. Some families, such as the histones, have genes with nearly identical sequences which are arranged tandemly along the chromosomal DNA and which function simultaneously to synthesize their gene products. Members of other gene families, however, though clustered on the chromosome, are not identical; such genes encode closely related proteins. Embryonic and adult globin polypeptides are examples of such a gene family. Within the globin gene cluster are also found "globinlike" sequences that are nonfunctional because of deletions in essential regulatory regions. These segments of DNA may represent vestiges of globin genes that were once functional but, during evolution, have lost their physiological role.

Polypeptide Production

Genetic engineering methodology has provided means for the production of potypeptides and proteins. Although this expression of heterologous genes is a function of a variety of complex factors, maximizing the expression of cloned sequences has been under intense and rapid development. It is now possible to produce a wide variety of foreign proteins in *E. coli.* These range from enzymes useful in molecular biology to a vast range of polypeptides with potential human therapeutic applications, such as insulin, interferon, growth hormone, immunoglobins, and enzymes involved in the dynamics of blood coagulation.

Biotechnology on a large scale has evolved as a result of genetic engineering in the laboratory. Human insulin manufactured through the use of genetically engineered *E. coli* has been used safely in diabetes therapy. Other polypeptides such as human growth hormone and interferon have been in production. Research has been under way to manufacture vaccines and many other therapeutic and diagnostic agents by using the new technology.

Medical Applications

The screening of appropriately restricted human DNA with suitable molecular probes, that is, true genes or simply nonrepeated sequences of human DNA cloned in a vector, has disclosed differences between individuals with respect to the size of the restricted DNA fragment containing DNA sequences homologous to the probe. These types of variants within a population, called restriction-fragment-length polymorphisms, can be used to map the human genome.

Restriction-fragment-length polymorphisms are inherited by mendelian segregation and are distributed in populations as classical examples of common genetic polymorphisms. If such a DNA variant is located dose to a defective gene (which cannot be tested for directly), the DNA variant can be used as a marker to detect the presence of the disease-causing gene. It is estimated that detection of 150 to 300 different DNA markers of this type distributed randomly throughout the human genome would allow detection of any disease-producing gene. Such information would be important in determining individual susceptibility to diseases and in taking measures to prevent them.

In cases where the specific mutation in DNA has been identified, such as sickle cell anemia, certain restriction enzymes that recognize the abnormal DNA sequence at the mutant site can be used to demonstrate the mutation without study of the affected and unaffected family members (Fig. 3.3). By using restriction enzymes in molecular hybridization experiments, changes in DNA sequences that occur in other hemoglobinopathies (genetically determined defects of function or synthesis of hemoglobins) may also be recognized directly by using the appropriate probes. Such changes include deletions, mutations of restriction sites, and deletions of restriction sites. This kind of information provides means of antenatal diagnosis of human diseases by analysis of fetal blood samples.

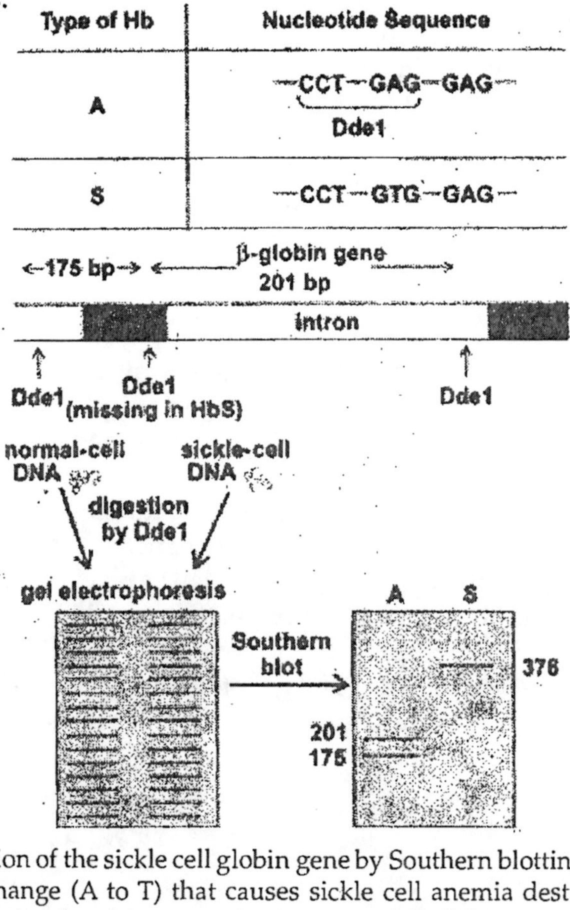

Fig. 3.3. Detection of the sickle cell globin gene by Southern blotting. The base change (A to T) that causes sickle cell anemia destroys a Dde 1 site that is present in the normal β-globin gene. This difference can be detected by digesting sickle cell and normal DNA with Dde 1 and performing Southern blot hybridization. Normal DNA will produce two Dde1 fragments of 201 and 175 base pairs, whereas sickle cell DNA will produce a single fragment of 376 base pairs.

A number of human diseases are caused by the abnormal function of a major single gene. Various thalassaemias--conditions characterized by reduced or absent synthesis of alpha- or beta-globin chains (the protein components of hemoglobin)--are examples of such diseases. If a normal copy of the defective gene could be introduced and was expressed in the appropriate cells of an individual suffering from such a genetic disorder, the normal gene product would be able to alleviate the problems associated with the presence of the defective gene. Since hemoglobin is produced by certain bone marrow cells that can easily be retrieved, thalassaemia might be treated by the addition of normal isolated genes (in the form of DNA) to the individual's abnormal erythropoietic marrow cells. After the normal genes have been taken up, the manipulated marrow cells could be returned to the person for proliferation and production of normal hemoglobin.

Genetic Modification of Plants

Experiments showing the successful transfer and expression of foreign DNA in plant cells using the Ti plasmid, as well as the demonstration that whole plants can be regenerated from cells containing mutated regions of T-DNA, indicate that the Ti plasmid system may be an important tool in the genetic engineering of plants. Such a system will help in the identification and characterization of plant genes as well as provide basic knowledge about gene organization and regulation in higher plants. Once genes useful for crop improvement have been identified, cloned, and stably inserted into the plant genome, it may be possible to engineer plants to be resistant to environmental stress, to pests, and to pathogens.

4

Molecular Biology

Introduction

It is the study of structural and functional properties of biological systems, pursued within the context of understanding the roles of the various molecules in living cells and the relationship between them. Historically, molecular biology evolved as a biological discipline with roots derived from biophysics, genetics, and biochemistry. Since its inception, a prime focus of the field has been the molecular basis of genetics, and with the demonstration in the mid-1940s that deoxyribonucleic acid (DNA) is the genetic material, emphasis has been on structure, organization, and regulation of genes. Initially, molecular biologists restricted their studies to bacterial and viral systems, largely because of the systems' genetic and biochemical simplicity. However, a series of conceptual and technological developments occurred rapidly during the late 1970s that permitted molecular biologists to approach a broad spectrum of plant and animal cells with experimental techniques. One of the major factors that rendered higher plant and animal cells amenable tc molecular analysis has been the development and applications of genetic engineering. The implementation of recombinant DNA technology allowed the isolation and selective modification of specific genes, thereby reducing both their structural and functional complexity and facilitating the study of gene expression in higher cells.

Systems and Approaches

Although molecular biology initially developed as an independent discipline, the concepts and techniques used by molecular biologists have been rapidly and effectively employed to resolve numerous cellular, biological, and biochemical problems—becoming routine at both the basic and applied levels. The result is that molecular biology has emerged as a high-resolution approach to solving questions previously addressed in more traditional biochemical terms.

Intact Cells and Cell-Free Systems

While the ultimate objective of molecular biology is to understand biological function and its regulation at the molecular level, the complexity of intact higher organisms has in most cases precluded direct analysis. It has therefore been necessary to pursue the characterization and functional properties of biologically important molecules in a series of simplified systems. Since the 1960s many such investigations have been carried out using intact cells. *Escherichia coli* has been extensively examined because of its limited number of cellular functions and the corresponding restricted amount of genetic information encoded in the bacterial chromosome. Simple eukaryotic cells, such as protozoa and yeast, offer similar advantages and have also been studied. For these same reasons, bacteriophage and animal viruses have provided molecular biologists with the ability to study the structural and functional properties of molecules in intact cells.

The development of techniques for propagation of higher eukaryotic cells, including human cells, in vessels has permitted assessment of molecular function in living cells under conditions far simpler than those that exist in intact organisms. Although analysis of molecular events and regulatory mechanisms is complicated by the presence of the cell's complete set of genes and by a multitude of complex cellular functions, the ability to work with a single cell type under defined conditions has proved to be extremely advantageous. Cell culture is a widely utilized approach that has become important with advances in the understanding of specific cell growth requirements and of factors necessary for various differentiated functions.

Cell-free systems, in which components of cells are isolated and the properties of specific molecules and biochemical reactions are investigated in test tubes, have advanced the understanding of complex cellular events. Systems have been developed that will carry out a number of biologically essential molecular processes such as DNA synthesis, RNA synthesis, and protein synthesis. Most important, these systems allow the specific molecules participating in the processes to be identified and characterized, frequently by addition, subtraction, or modification of the potentially relevant component. The ability to carry out biologically important reactions in cell-free systems reduces the number of variables and simplifies interpretation of experimental results. Yet, one must accept conclusions from experiments on such systems with the proviso that conditions in a test tube are somewhat artificial.

Gene Mapping

The recognition of DNA as the genetic material coupled with the discovery that genes reside in chromosomes resulted in an intensive effort to map genes to specific chromosomes. Initially genes were assigned to chromosomes on the basis of correlations between modifications in cellular function, particularly biochemical defects, and the addition, loss, or modification of specific chromosomes. These early efforts at chromosomal location of genetic traits were largely restricted to lower animals and plants since they have relatively few chromosomes. Chromosome analysis was histochemical, that is, it was done by staining metaphase chromosomes spread on microscope slides and examining the preparations by light microscopy. Despite the limited resolution of this approach, chromosomal assignments of a number of genes were made. In fact, in the absence of knowledge of the biochemical or genetic basis for several human disorders such as Down syndrome, a series of chromosomal aberrations permitted identification of chromosomes encoding genes related to these diseases. Another approach which, early on, permitted assignment of genes to chromosomes was introduction of mutations by radiation or chemicals followed by a correlation between aberrations in metaphase chromosome morphology and biochemical or cellular abnormalities.

A major breakthrough in mapping genes was the development

of somatic cell genetics. This is an approach in which, for example, human and hamster cells are fused, resulting in a hybrid cell initially containing the complement of human and hamster chromosomes. As the cells grow and divide in culture, the hamster chromosomes are retained while there is a progressive loss of human chromosomes. By correlating the loss of human biological or biochemical traits with the loss of specific human chromosomes, a number of human genes have been successfully mapped.

The development of methods for isolating genes and for determining the genetic sequences of the DNA in which the genes are encoded, led to rapid advances in gene mapping at several levels of resolution. Localization of specific genes to chromosomes is routinely carried out with cloned genes as probes. For example, isolated human genes are radioactively labeled and used to recognize and interact with an identical copy of the human gene in DNA isolated from human-hamster hybrid cells. Such analysis of a series of human-hamster hybrids, each hybrid containing limited numbers of various human chromosomes, offers an effective method for chromosome assignment. Further information about the segment of a chromosome in which a specific gene resides can be obtained by directly determining the DNA sequences of both the gene itself and the surrounding region.

Chromosome localization of specific genes has numerous applications at both the basic and clinical levels. At the basic level, knowledge of the positions of various genes provides insight into potentially functional relationships. At the clinical level, chromosome aberrations are now routinely used in prenatal diagnosis of an extensive series of human genetic disorders, and several chromosomal modifications have been linked to specific types of cancer. Knowledge of genetic defects at the molecular level has permitted the development of diagnostic procedures that in some instances, such as sickle cell anemia, are based on a single nucleotide change in the DNA.

Recombinant DNA

Recombinant DNA technology has provided molecular biology with an extremely powerful tool. Though introduced only in the mid-1970s, genetic engineering has had a major impact on science, medicine, technology, and society in general, providing numerous opportunities for improvement of the quality of human existence. In broad terms, applications of recombinant DNA technology can be divided into four areas--biomedical, basic biological, agricultural, and industrial. Biomedical applications include the elucidation of the cellular and molecular bases of a broad spectrum of diseases, as well as both diagnostic and therapeutic applications in clinical medicine.

In a strictly formal sense, the term recombinant DNA designates the joining or recombination of DNA segments. However, in practice, recombinant DNA has been applied to a series of molecular manipulations whereby segments of DNA are rearranged, added, deleted, or introduced into the genomes of other cells. The processes associated with modifying DNA molecules are extremely precise, permitting addition, subtraction, or alterations of specific genes or of particular regions of genes.

The importance of recombinant DNA can be readily appreciated because of the pivotal role of genetic sequences in defining the structural and functional properties of cells, tissues, and organs. Selective expression of particular genes provides the unique properties characteristic of specialized cells, for example, hemoglobin production by erythropoietic cells and myosin production by muscle cells. A progressive expression of certain genes coupled with selective repression of others is necessary for the successful orchestration of development and differentiation. Prerequisite for maintaining the physiological integrity of cells, tissues, and organs is the capacity to modulate gene expression in response to changes in cellular requirements: for example, activation of genes associated with cell proliferation in conjunction with tissue regeneration, or modifications in the expression of genes encoding metabolic enzymes in association with fluctuations in nutrient levels.

Equally important are the findings that numerous diseases, including cancer and disorders that involve inborn errors of metabolism, are related to a variety of genetic alterations, including rearrangements or mutations in specific DNA sequences. It therefore follows that by having the capacity to manipulate the organization of genes, the capacity to correct genetic defects is not an unrealistic expectation. And while the latter goal requires technological developments and ethical justifications, the available repertoire of genetic manipulations has enabled scientists to detect several clinically important genetic defects in the fetus (such as sickle-cell anemia) and to produce biologically important molecules in cell-free systems on a scale which allows them to be used for clinical treatment (for example, insulin, growth hormone, and interferon).

Techniques

Historically, the ability to manipulate or "engineer" genetic sequences is based on several developments, as discussed in the following sections.

1. **Methods for breaking and rejoining DNA.** The precise breaking and rejoining of DNA has been made possible by the discovery of restriction endonucleases, enzymes that have the ability to recognize specific DNA sequences and to cleave the double helix precisely at these sites. Also important is the ability to join fragments of DNA together with the enzyme DNA ligase, and the techniques to determine the nucleotide sequence of genes and thereby confirm the identity and location of structural and regulatory sequences.

2. **Carriers for genetic sequences.** Bacterial plasmids, that is, circular double-stranded DNA molecules that replicate extrachromosomally, have been modified so that they can serve as efficient carriers for segments of DNA, complete genes, regions of genes, or sequences contained within several different genes. Bacteriophage (viruses which infect and replicate in bacteria) and animal viruses, such as SV40 (a simian virus), retroviruses (RNA tumor viruses), and bovine papilloma virus, have also been successfully utilized as DNA carriers. In the vernacular of a genetic engineer, these carriers are referred to as cloning vectors. Host cells in which vectors containing cloned genes can replicate range from bacteria to numerous other cells, including normal, transformed, and malignant human cells.

3. **Introduction of recombinant DNA molecules.** Genetic sequences in the form of isolated DNA fragments, or chromosomes, or of DNA molecules cloned in plasmid

vectors can be introduced into host cells by a procedure referred to as transfection or DNA-mediated gene transfer--a technique that renders the cell membrane permeable by a brief treatment with calcium phosphate, thereby facilitating DNA uptake. Genes cloned in viruses can also be introduced by infection of host cells.

4. **Selection of cells containing cloned sequences.** Several approaches have been developed for identification of bacterial, plant, or animal cells containing cloned genetic sequences. Bacterial cells containing plasmids with cloned genes can be detected by selective resistance or sensitivity to antibiotics. In addition, the presence of introduced genes in bacterial, plant, or animal cells can be assayed by a procedure known as nucleic acid hybridization, in which a radioactive probe consisting of a region of a cloned gene is permitted to anneal with complementary sequences in the host cells. The recognition of the complementary sequence in the host cell is reflected by formation of a sequencespecific hybrid.

5. **Amplification.** Amplification of genetic sequences cloned in bacterial plasmids is efficiently achieved by treatment of host cells with antibiotics which suppress replication of the bacterial chromosome, yet do not interfere with replication of the plasmid with its cloned gene. Sequences cloned in bacterial or animal viruses are often amplified by virtue of the ability of the virus to replicate preferentially.

6. **Expression.** Expression of cloned human genes can be mediated by regulatory sequences derived from the natural gene, from exogenous genes, or by host cell sequences. Selection of regulatory sequences affixed to the cloned gene is based on the host cell or the regulatory signal to which expression is desired. For example, production of large quantities of a protein or RNA from a cloned human gene can often be achieved in a bacterial cell if the human gene is under the control of bacterial regulators and regulatory sequences. If a gene is introduced into a human cell to compensate for a genetic

defect, however, responsiveness to normal physiological signals is an absolute necessity.

Applications

Two clinically important genes, human insulin and human growth hormone, have been cloned and introduced into bacteria under conditions where biologically active hormones can be produced. It would be difficult to underestimate the clinical relevance of cloned human insulin and growth hormone genes. Availability of biosynthetic human insulin will permit management of those patients with diabetes who cannot be successfully treated with bovine or porcine insulin and will ensure a continuous supply of insulin for everyone with diabetes. The unlimited availability of biosynthetic human growth hormone from cloned genes will effectively eliminate the shortage of this hormone and will make it possible for all children with growth hormone deficiency to be treated.

Progress has been made in applications of recombinant DNA technology to the resolution of agricultural problems, especially for the improvement of both crops and livestock. For example, there are possibilities for nitrogen fixation by crops which presently deplete the soil of nutrients, for enhancement of photosynthetic capabilities, and for improvement of the nutritional levels of storage protein in grains.

Biophysical Analysis

Understanding of the structural properties of molecules and the interaction between molecules that constitute biologically important complexes has been facilitated by biophysical analysis. For example, developments in the resolution offered by techniques such as electron microscopy, x-ray diffraction, and neutron scattering have provided valuable insight into the structure of chromatin, the protein-DNA complex which constitutes the genome of eukaryotic cells. These techniques have also provided clues about modifications in chromatin structure that accompany functional changes. One possible application of biophysical analysis is the diagnosis of human disorders by adaptation of nuclear magnetic resonance for tissue and whole body evaluation of soft tissue tumors, blood flow, and cardiac function.

Flow of Molecular Information

Information for all cellular activities is encoded in DNA; selective elaboration of this information is prerequisite to meeting both structural and biochemical requirements of the cell. In this regard, there are three major areas of investigations by molecular biologists: (1) the composition, structure, and organization of chromatin, the protein-DNA molecular complex in which genetic information is encoded and packaged; (2) the molecular events associated with the expression of genetically encoded information so that specific cellular biochemical requirements can be met; and (3) the molecular signals that trigger the expression of specific genes and the types of communication and feedback operative to monitor and mediate gene control.

Gene Composition and Organization

In both prokaryotic and eukaryotic cells, genetic information is contained in DNA molecules, but there are differences in the composition and organization of genes which are directly related to variations in expression and control. In all cases, cellular proteins are encoded in DNA sequences utilizing a triplet code of four deoxynucleotide bases (adenine, guanine, thymine, and cytosine), the building blocks of DNA. The DNA sequences also provide for major elements of secondary and possibly higher-order structural features of genes and serve as recognition sites for molecules that participate in gene structure and function. Typically, the central component of the gene is a series of nucleotides encoding a protein, and these protein coding sequences are flanked by stretches of nucleotides which influence expression.

Deoxyribonucleic acid is a long polymer with two polynucleotide chains coiled around a central axis in the configuration of a double helix. The backbone or constant component of each nucleotide chain consists of deoxyribose sugar moieties linked covalently by 3′,5′-phosphodiester bonds. The variable component of DNA which renders the molecule capable of encoding genetic information is its sequence of four nitrogenous bases: two purines (adenine and quanine) and two pyrimidines (cytosine and thymine). The bases are covalently linked to the deoxyribose moiety of the backbone by N-glycosidic bonds. The bases of the two nucleotide chains are linked to each other by weak hydrogen bonds, with adenine always paired with thymine and guanine with cytosine. The two polynucleotide chains are antiparallel (the 3′ end of one strand faces the 5′ end of the other) and oriented so that the purines and pyrimidines reside within the helix and the sugar-phosphate backbone is on the outside. The viability of the double-helical model for DNA has been due to its compatibility with two key molecular and biochemical properties of the molecule: the ability to replicate, and the ability to elaborate genetic information by serving as a template for the synthesis of complementary messenger ribonucleic acid (RNA) molecules which then act as template for the synthesis of genetically encoded proteins.

The representation, organization, and packaging of genes is directly related to their expression; hence, much ongoing effort has been directed toward studying these properties of both prokaryotic and eukaryotic genes. In the bacterium *E. coli* there is a single chromosome containing a circular DNA molecule approximately 0.06 in. (1.5 mm) in length when unfolded, and with about 4×10^6 nucleotides encoding a limited number of genes (approximately 4000), most represented only once. In contrast, human cells contain 23 pairs of chromosomes, each with a linear DNA molecule. Several yards of DNA are found in the nucleus of every human somatic cell, encoding many thousands of genes, some represented only once per set of chromosomes and others represented by up to several hundred copies.

Protein-DNA Complexes

Both prokaryotic and eukaryotic chromosomes exist as protein-DNA complexes. Although the representation of proteins in prokaryotic chromosomes is limited, several bacterial proteins have been extensively characterized and shown to interact with specific regions of genes, thereby regulating expression. Lac repressor is an example of a bacterial protein which interacts with a specific series of nucleotides in the gene encoding b-galactosidase and influences the extent to which b-galactoside messenger RNA (mRNA) can be synthesized.

In contrast, the chromosomes of eukaryotic cells are protein-DNA complexes with a 2:1 protein-to-DNA ratio. A large body of experimental results indicates that the chromosomal proteins of eukaryotic cells are responsible for packaging the huge DNA molecule within the confines of the cell nucleus, which is only several micrometers in diameter. Modifications in the interactions of chromosomal proteins with eukaryotic DNA have also been shown to occur in conjunction with changes in gene activity and have therefore been implicated in the control of gene expression. Accessibility of specific regions of genes to nucleases, enzymes which degrade DNA unprotected by chromosomal proteins, has provided an effective probe for establishing such structural-functional relationships.

There are five classes of basic chromosomal proteins known as histories which are enriched in arginine, lysine, and histidine residues. These histone proteins are present in an amount equivalent to DNA and have been shown to interact with one another and with DNA to form the primary unit of eukaryotic chromatin structure, the nucleosome. The other chromosomal proteins of eukaryotic cells are designated nonhistone proteins and constitute a complex and heterogeneous array of molecules which interact with DNA and appear to participate in structural, enzymatic, and regulatory activities.

DNA Rearrangements

It has become evident that genes are not necessarily static. Alterations in the representation and organization of DNA nucleotide sequences can occur in both bacteria and eukaryotic cells. For example, viruses can introduce DNA sequences into chromosomes, and a series of transpositions or movements of DNA sequences from one location to another has been documented in bacterial, plant, and animal cells. Flexibility in gene structure and organization appears to be related to biological function. There are several biological situations in which rearrangements of DNA sequences occur: (1) in conjunction with production of immunoglobulins; (2) with the development and onset of both solid tumors and human leukemias; and (3) with changes in the phenotypic properties of corn.

Extrachromosomal Genes

A complete evaluation of a cell's genes and gene functions must take into account the presence of extra chromosomal DNA.

Bacterial cells can carry extrachromosomal genes, such as those that confer antibiotic resistance, in the form of plasmids, which are covalently closed circular DNA molecules. Stable sets of extrachromosomal genes in eukaryotic cells are those in the mitochondria, and in chloroplasts, which encode several of the proteins required for the functional integrity of those organelles. Extrachromosomal DNA can also accumulate in eukaryotic cells in conjunction with the development of drug resistance, as has been shown for the amplification of dihydrofolate reductase genes as cells acquire tolerance to the folate antagonist methotrexate.

Gene Expression

Expression of genetic traits both in bacteria and in eukaryotic cells requires the selective elaboration and processing of genetically encoded information. This occurs by transcription of DNA into RNA molecules that are complementary to the gene; processing of these RNA transcripts into mRNAs; and association of the mRNAs with ribosomes where the nucleotide sequences serve as a template for synthesis of a protein. Three types of RNA are required for the process: mRNA which carries the blueprints for proteins from DNA to ribosomes; ribosomal RNAs (rRNA) which serve as major components of the protein synthesis apparatus; and transfer RNAs (tRNA) which carry amino acids, the primary units of proteins, to the ribosome-associated mRNAs where they are Assembled into proteins.

In prokaryotic cells, gene expression is rapid and considerably less complex than in eukaryotic cells. Prokaryotic mRNAs are generally encoded in contiguous DNA nucleotides, and with minimal modifications they function as templates for protein synthesis. In most situations, prokaryotic mRNAs are short-lived and some are polycistronic, that is, several proteins are encoded in a single mRNA molecule regulated by a single set of control sequences.

Eukaryotic mRNAs in general are not encoded by contiguous DNA nucleotides; rather, a precursor RNA which contains encoded segments of the mRNA (exons) separated by intervening sequences (introns) is transcribed from the DNA. The intervening sequences are excised and the protein coding sequences are joined together, or "spliced." Further aspects of eukaryotic mRNA processing that occur within the nucleus involve enzymatic modifications of the 3′ terminus [addition of a series of adenosine monophosphate (AMP) residues], the 5′ terminus (addition of a methylguanine-containing moiety referred to as a CAP structure), and various bases within the mRNA molecule (for example, methylation). During this processing the mRNA becomes complexed with proteins, forming a ribonucleoprotein particle which is exported to the cytoplasm where it can associate with ribosomes. , These complicated eukaryotic mRNA processing steps require from 20 to 45 min; present understanding of the specific molecular mechanisms involved is minimal. The diversity of eukaryotic mRNAs is reflected by the proteins they template, the manner in which they are processed, and their lifetimes which range from several minutes to extremely long periods of time and vary as a function of the biological situation.

Protein synthesis in both prokaryotic and eukaryotic cells is a series of critically orchestrated processes in which the information genetically encoded in the mRNA is utilized for the step-by-step covalent assembly of amino acids into a protein molecule. The process occurs entirely in the cytoplasm of eukaryotic cells after completion of mRNA processing, while in prokaryotic cells protein synthesis can be initiated during completion of mRNA synthesis. Protein synthesis requires availability of ribosomes, tRNAs, and amino acids, along with a series of enzymes, factors, and energy sources. In many instances, after completion of protein synthesis, assembly of protein subunits or enzymatic modifications such as addition of acetate, phosphate, methyl, carbohydrate, or polyadenosine diphosphate-ribose moieties are required to yield a biologically functional protein molecule.

Regulation of Gene Expression

The mechanisms operative in the selective expression of specific genes and those responsible for the inactivity of others are major issues in molecular biology. Attention has been directed primarily toward identification of nucleotide sequences that are involved in regulating gene transcription and toward characterization of molecules that influence the availability of genes for transcription and those that affect the processing of RNA transcripts. In prokaryotic cells, gene expression is largely mediated at the transcriptional level, that is, the extent to which many bacterial proteins are synthesized is directly related to the extent that the gene is transcribed into mRNAs. Yet, posttranscriptional control is also an element in prokaryotic gene regulation, since the extent to which some bacterial proteins are synthesized is governed by the ability of mRNAs to serve as templates for protein synthesis or by the efficiency of specific biochemical steps which participate in the protein synthetic process.

In eukaryotic cells, regulation of gene expression is both transcriptionally and posttranscriptionally mediated. In those situations where expression is controlled at the transcriptional level, there appear to be similarities in the mechanisms operative in prokaryotic and eukaryotic cells. However, because of the numerous steps associated with the nuclear and cytoplasmic processing of eukaryotic mRNAs there are many potential targets where posttranscriptional control can occur. Data suggest that in normal biological situations processing of mRNA transcripts within the cell nucleus and the ability of cytoplasmic mRNAs to associate with ribosomes and serve as functional templates for protein synthesis can influence the expression of specific genes. Also, the molecular basis of several human diseases probably resides in defects at various posttranscriptional levels of gene expression. For example, in at least one of the b-thalassemias, a human blood disorder in which there is a reduced representation of b-globin, an impairment exists in the splicing of the b-globin mRNA precursor molecule.

An area of primary relevance to understanding regulation of gene expression, and one where knowledge is extremely limited, is the identification and characterization of cellular and molecular signals which monitor the requirements for specific proteins and which consequently activate the molecular mechanisms required to modulate the transcription, processing, or turnover of the respective mRNAs. Considerable attention has been directed toward the involvement of molecules such as steroid hormones, polypeptide hormones, and cyclic nucleotides as mediators of gene expression. The concept of autogeneous control has also been considered as a mechanism that in some situations can provide the molecular basis for regulating expression of a particular gene by using the gene product to destabilize its mRNA or to block further transcription of the gene.

5

Enzyme

Introduction

It is a catalytic protein produced by living cells. The chemical reactions involved in the digestion of foods, the biosynthesis of

Some representative enzymes, their sources, and reaction specificities

Enzyme	*Some sources*	*Reaction catalyzed*
Pepsin	Gastric juice	Hydrolysisofproteinstopeptidesandaminoacids
Urease	Jackbean, bacteria	Hydrolysisofureatoammoniaandcarbondioxide
Amylase	Saliva, pancreatic juice	Hydrolysisofstarchtomaltose
Phosphorylase	Muscle, liver, plants	Reversible phosphorolysis of starch or glycogen to glucose-1-phosphate
Transaminases	Many animal and plant tissues	Transfer of an amino group from an amino acid to a keto acid
Phosphohexose isomerase	Muscle, yeast	Interconversion of glucose-6-phosphate and fructose-6-phosphate
Pyruvic carboxylase	Yeast, bacteria, plants	Decarboxylation of pyruvate to acetaldehyde and carbon dioxide
Catalase	Erythrocytes, liver	Decomposition of hydrogen peroxide oxygen and water
Alcohol dehydrogenase	Liver	Oxidationofethanoltoacetaldehyde
Xanthine - oxidase	Milk, liver	Oxidationofxanthineandhypoxanthinetouricacid

macromolecules, the controlled release and utilization of chemical energy, and other processes characteristic of life are all catalyzed by enzymes. In the absence of enzymes, these reactions would not take place at a significant rate. Several hundred different reactions can proceed simultaneously within a living cell, and the cell contains a comparable number of individual enzymes, each of which controls the rate of one or more of these reactions. The potentiality of a cell for growing, dividing, and performing specialized functions, such as contraction or transmission of nerve impulses, is determined by the complement of enzymes it possesses. Some representative enzymes, their sources, and reaction specificities are shown in the Table 5.1.

Characteristics

Enzymes can be isolated and are active outside of the living cell. They are such efficient catalysts that they accelerate chemical reactions measurably, even at concentrations so low that they cannot be detected by most chemical tests for protein. Like other chemical reactions, enzyme-catalyzed reactions proceed only when accompanied by a decrease in free energy; at equilibrium the concentrations of reactants and products are the same in the presence of an enzyme as in its absence. An enzyme can catalyze an indefinite amount of chemical change without itself being diminished or altered by the reaction. However, because most isolated enzymes are relatively unstable, they often gradually lose activity under the conditions employed for their study.

Chemical Nature

All enzymes are proteins. Their molecular weights range from about 10,000 to more than 1,000,000. Like other proteins, enzymes consist of chains of amino acids linked together by peptide bonds. An enzyme molecule may contain one or more of these polypeptide chains. The sequence of amino acids within the polypeptide chains is characteristic for each enzyme and is believed, to determine the unique three-dimensional conformation in which the chains are folded. This conformation, which is necessary for the activity of the enzyme, is stabilized by interactions of amino acids in different parts of the peptide chains with each other and with the surrounding medium. These interactions are relatively weak and may be disrupted readily by high temperatures, acid or alkaline conditions, or changes in the polarity of the medium. Such changes lead to an unfolding of the peptide chains (denaturation) and a concomitant loss of enzymatic activity, solubility, and other properties characteristic of the native enzyme. Enzyme denaturation is sometimes reversible. It is possible, in some cases, to obtain conditions under which the denatured enzyme refolds into its native conformation and regains its original catalytic activity.

Many enzymes contain an additional, nonprotein component, termed a coenzyme or prosthetic group. This may be an organic molecule, often a vitamin derivative, or a metal ion. The coenzyme, in most instances, participates directly in the catalytic reaction. For example, it may serve as an intermediate carrier of a group being transferred from one substrate to another. Some enzymes have coenzymes that are tightly bound to the protein and difficult to remove, while others have coenzymes that dissociate readily. When the protein moiety (the apoenzyme) and the coenzyme are separated from each other, neither possesses the catalytic properties of the original conjugated protein (the holoenzyme). By simply mixing the apoenzyme and the coenzyme together, the fully active holoenzyme can often be reconstituted. The same coenzyme may be associated with many enzymes which catalyze different reactions. It is thus primarily the nature of the apoenzyme rather than that of the coenzyme which determines the specificity of the reaction.

The complete amino acid sequence of several enzymes has been determined by chemical methods. By x-ray crystallographic methods even the exact three-dimensional molecular structure of a few enzymes has been deduced.

Classification and Nomenclature

Enzymes are usually classified and named according to the reaction they catalyze. The principal classes are as follows.

Oxidoreductases

These enzymes, which catalyze reactions involving electron transfer, play an important role in cellular respiration and energy production. Some of them participate in the process of oxidative phosphorylation, whereby the energy released by the oxidation of carbohydrates and fats is utilized for the synthesis of adenosine triphosphate (ATP) and thus made directly available for energy-requiring reactions. Most oxidoreductases require prosthetic groups or coenzymes, which participate in the catalytic reaction by serving as intermediate electron carriers. These coenzymes include the flavin nucleotides, the pyridine nucleotides, heme, and various metal ions, such as iron, copper, and molybdenum. Some oxidoreductases, the oxygenases, catalyze the direct incorporation of oxygen into their substrate. Oxidases are enzymes which utilize molecular oxygen as an electron acceptor, while dehydrogenases remove hydrogen atoms from their substrates and transfer them to an acceptor other than oxygen. For example, glucose oxidase catalyzes the reaction shown in reaction (1). This enzyme utilizes flavin adenine dinucleotide, a derivative of the vitamin riboflavin, as coenzyme.

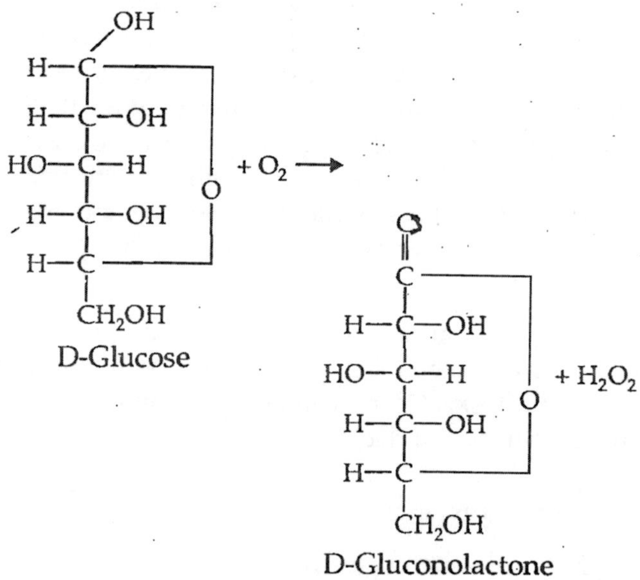

(1)

Transferases

These enzymes catalyze the transfer of a particular chemical group from one substance to another. Thus, transaminases transfer amino groups, transmethylases transfer methyl groups, and so on. An important subclass of this group are the kinases, which catalyze the phosphorylation of their substrates by transferring a phosphate group, usually from ATP, thereby activating an otherwise metabolically inert compound for further transformations. The reaction catalyzed by transaminases may be represented by reaction (2).

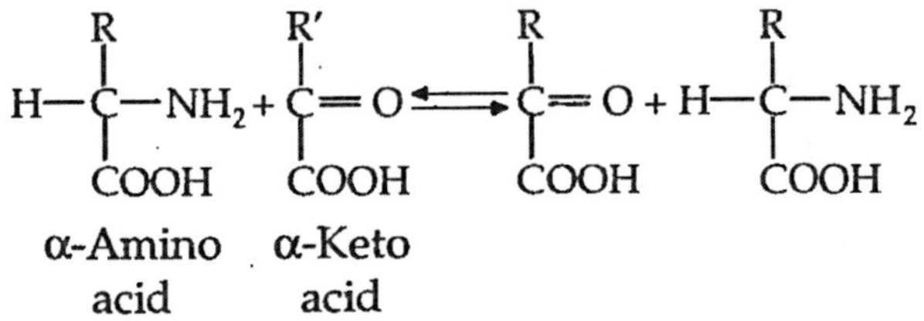

(2)

These enzymes require pyridoxal phosphate or pyridoxamine phosphate, derivatives of vitamin B_6, as coenzymes.

Hydrolases

These are enzymes which catalyze the hydrolysis of proteins (proteinases and peptidases), of nucleic acids (nucleases), of starch (amylases), of fats (lipases), of phosphate esters (phosphatases), and of other substances. Many hydrolases are secreted by the stomach, pancreas, and intestine and are responsible for the digestion of foods. Others participate in more specialized cellular functions. For example, cholinesterase, which catalyzes the hydrolysis of acetylcholine, as shown in reaction (3.), plays an important role in the transmission of nervous impulses.

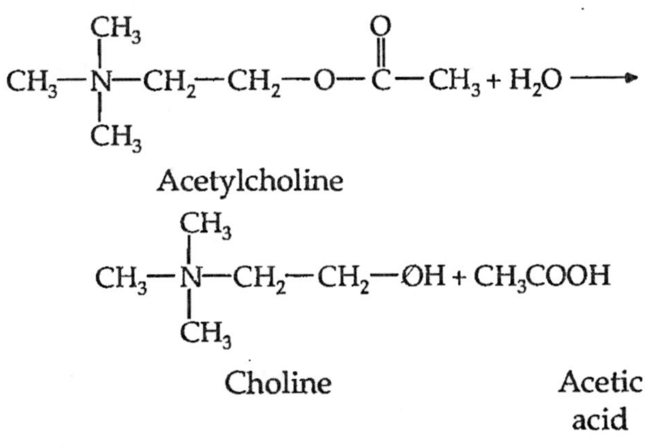

(3)

Lyases

These enzymes catalyze the nonhydrolytic cleavage of their substrate with the formation of a double bond. Examples are decarboxylases, which remove carboxyl groups as carbon dioxide, and dehydrases, which remove a molecule of water. The reverse reactions are catalyzed by the same enzymes, but are difficult to demonstrate in some cases because of an unfavorable equilibrium.

Thus pyruvic decarboxylase, which requires thiamine pyrophosphate and magnesium ions for activity, catalyzes the reaction that is shown in reaction (4).

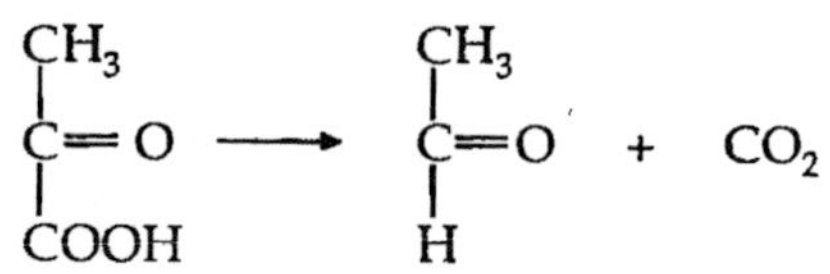

(4)

Pyruvic acid Acetaldehyde

Isomeruses

These are enzymes which catalyze the interconversion of isomeric compounds. For example, triose phosphate isomerase catalyzes the reaction shown in reaction (5).

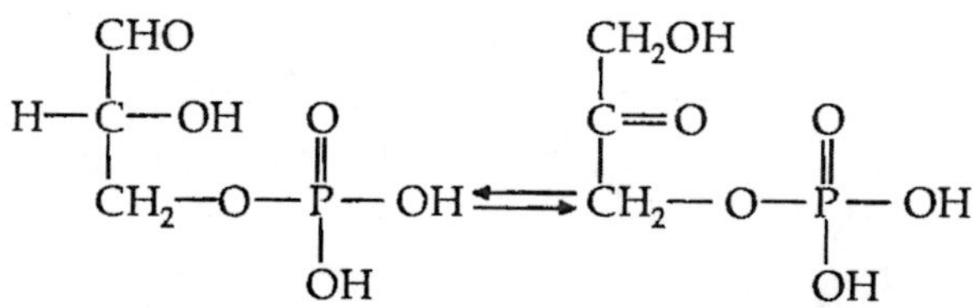

(5)

D-Glyceraldehyde-3-phosphate Dihydroxyacetone phosphate

Ligases or Synthetases

These enzymes catalyze endergonic syntheses coupled with the exergonic hydrolysis of ATP. They allow the chemical energy stored in ATP to be utilized for driving reactions uphill. An example is glutamine synthetase, which catalyzes the reaction shown in reaction (6).

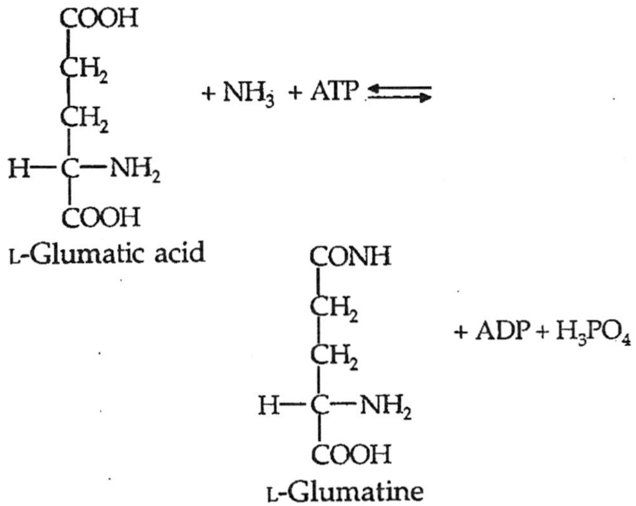

(6)

This enzyme, which is abundant in brain tissue, requires magnesium or manganese ions for activity.

Active Site

Since enzymes are large molecules and their substrates are often of low molecular weight, it appears probable that only a small portion of the enzyme protein, the active site, comes into contact with the substrate and is directly involved in catalyzing the reaction. The active site consists of a few amino acid residues, not necessarily adjacent to each other on the polypeptide chain but brought into proximity by the folding of the chain or chains. The active site also includes any coenzyme which may be required for activity. The function of the remainder of the protein molecule may be primarily to hold the components of the active site in the proper relative position and orientation. Part of the active site is involved in binding the substrate, while another part is responsible for the making or breaking of chemical bonds. There is much evidence to indicate that, for some enzymes, the binding of the substrates produces a conformational change which brings some reactive group of the enzyme in the proper position for the reaction to take place.

Because of their special environment at the active site of enzymes, coenzymes and certain amino acid residues acquire unique properties and reactivities which they do not possess in the free state. For example, when the proteinase chymotrypsin is treated with diisopropylfluorophosphate, one serine residue at the active site reacts rapidly with this reagent, resulting in complete inactivation of the enzyme. The other 27 serine residues of the protein do not react, and no reaction takes place when the protein molecule is first unfolded, that is, when it is denatured. On the other hand, amino acid residues remote from the active site of an enzyme may often be removed or chemically modified without affecting the catalytic activity.

Intracellular Organization

Enzymes are not distributed uniformly throughout the cell but are localized within various subcellular structures or compartments. For example, the enzymes responsible for glycolysis, the degradation of glucose to pyruvate, are located in the soluble, nonparticulate portion of the cell, whereas the enzymes that catalyze the further oxidation of pyruvate to carbon dioxide and water and couple this oxidation to the production of ATP are localized in specialized particles, the mitochondria. Similarly, while the activation of amino acids for protein synthesis takes place in free solution, the assembly of the amino acids into proteins takes place on a particle, the ribosome. Other enzymes are localized on the cell membrane, in the nucleus, and in other subcellular particles.

Control of Enzyme Activity

In order to preserve the stability of its internal chemical environment, a living organism must possess the means of regulating the rate of its metabolic reactions. This is accomplished by regulating both the amount and the activity of the enzymes catalyzing these reactions. Some of the control mechanisms involved are very complex and not yet completely understood, although much progress has been made in this field. A few examples are mentioned below.

Proenzyme Activation

One way by which the activity of enzymes is controlled is as follows. The enzyme is produced in an inactive form, termed a zymogen or proenzyme, and converted into the active form only when needed. This is the case with thrombin, an enzyme which causes blood to clot by catalyzing the conversion of the soluble protein, fibrinogen, to the insoluble fibrin. Blood does not normally contain thrombin but does contain its inactive precursor, prothrombin. It is only when prothrombin is converted to thrombin (by another enzyme) that clotting results. The mechanism is shown in reaction (7).

Prothrombin
(inactive)
↓
Thrombin
(active)

$$\text{Fibrinogen} \xrightarrow{\hspace{2cm}} \text{Fibrin (clot)} \qquad (7)$$

Another example of enzyme activation occurs with the proteinase trypsin, which is stored and secreted by the pancreas as the inactive trypsinogen. In the intestine, trypsinogen is converted to trypsin by the enzyme enterokinase. The conversion of trypsinogen to trypsin is also catalyzed by trypsin itself, the activation being therefore autocatalytic. The activation involves the splitting of a peptide from one end of the trypsinogen molecule.

Feedback Inhibition

Other mechanisms of enzyme control are shown by reaction sequence

$$A \xrightarrow{E_1} B \xrightarrow{E_2} C \xrightarrow{E_3} D \qquad (8)$$

(8) which depicts a hypothetical metabolic pathway in which the starting material *A* is converted to the metabolite *D* required by the cell. One type of control, termed feedback inhibition, results from a specific, reversible, inhibition of enzyme E_1 by the end product *D*. This phenomenon is obviously of great advantage to the cell, since it allows the concentration of the product *D* to be accurately regulated: When *D* reaches a certain concentration, it decreases the rate of its own synthesis by preventing the conversion of *A* to *B*; the intermediates, *B* and *C*, do not accumulate; and the starting material, *A*, is spared for other reactions for which it may

be needed. As soon as the concentration of *D* decreases, the inhibition of enzyme E_1 is relieved and the synthesis of *D* resumes.

This type of control is also termed allosteric control, and the enzyme involved, an allosteric enzyme. The allosteric enzyme contains a specific binding site for the controlling metabolite, *D*, such that when *D* is bound to this site the enzyme is inhibited. Positive allosteric control, or allosteric activation, has also been observed. In this case, the binding metabolite, which may be the product of another pathway, causes activation of the enzyme. Some enzymes are susceptible to allosteric control, both positive and negative, by a number of different metabolities.

Induction and Repression

Controls also exist at the level of enzyme synthesis. One of these controls is enzyme induction. For example, if the inducible enzymes E_1, E_2, and E_3 are required to utilize substance *A*, the enzymes are not produced by the cell when *A* is not available. Only in the presence of an inducer, which may be *A* or a related compound, are enzymes E_1, E_2, and E_3 synthesized. Another, related, type of control is enzyme repression. In this case the control is exerted by the end product of a metabolic pathway, for example, compound *D*. Thus, if enzymes E_1, E_2, and E_3 are repressible enzymes, the accumulation of *D* prevents their synthesis by the cell. Enzyme repression and induction involve complex regulatory mechanisms which are under direct genetic control.

Purification

One of the prerequisites for a study of the structure and mechanism of action of an enzyme is its availability in pure form. The enzyme must therefore first be extracted from the cells or tissue in which it is found and separated from many proteins and other substances present in the crude extract. Because of the instability of most enzymes, however, their isolation requires special techniques, some of which are described below.

Cell Destruction

If the enzyme is present in cells, the cells must first be broken.

This can be accomplished by grinding with abrasives, by freezing and thawing, or by exposure to ultrasonic vibrations. Although many enzymes are readily released in soluble form by these methods, some remain associated with membranes or subcellular particles and may be difficult to solubilize.

Heating with Substrate

Some enzymes can be purified by taking advantage of the fact that they are stabilized in the presence of their substrate and can then be heated for a short time to a temperature at which less stable proteins are denatured and precipitated.

Selective Precipitation

Enzymes contain many ionizable groups, such as carboxyl groups, which may be negatively charged, and amino groups, which may be positively charged. Since the degree of ionization of each group depends on the pH of the solution, the total charge on the protein varies greatly with pH. The pH at which the protein has no net charge, its isoelectric point, is usually the pH of minimal solubility. Thus an enzyme can often be selectively precipitated from a solution by adjusting the pH to its isoelectric point.

Electrophoresis

An enzyme can also be separated from proteins with different electric charges by electrophoresis. This method consists of applying an electric field to the solution and allowing each protein to migrate in a direction and at a rate which depend on its charge.

Ammonium Sulphate Fractionation

Another useful method of separating proteins from each other is by ammonium sulphate fractionation. The solubility of proteins decreases as the concentration of ammonium sulphate is raised, and when the salt concentration is sufficiently high, the protein precipitates. Since different proteins precipitate at different ammonium sulphate concentrations, it is possible, by increasing the salt concentration in steps and collecting the precipitate that appears at each stage, to obtain considerable purification.

Dialysis

Salts and organic impurities of low molecular weight can be easily removed from enzymes by dialysis. This consists of placing the enzyme solution in a bag made from a semipermeable membrane, such as cellophane, and immersing it in water or in a buffer solution. The protein remains in the bag, while solutes of low molecular weight diffuse out. An enzyme may be separated from smaller or larger proteins by gel filtration chromatography or by preparative ultracentrifugation.

Other Techniques

Other methods frequently used for enzyme purification include precipitation at low temperatures with organic solvents, such as acetone or alcohol; selective adsorption on gels, such as calcium phosphate or alumina; and chromatography on ion-exchange resins or modified cellulose. For the complete purification of an enzyme, a combination of these methods must be employed. No generally applicable enzyme purification procedure exists. The most satisfactory method must be determined empirically for each enzyme.

The final step in the purification of many enzymes is crystallization. The first crystalline enzyme, urease, was prepared by J. B. Sumner in 1926, and since then many other enzymes have been crystallized.

Isozymes

It has been found that many enzyme preparations, even after crystallization, still consist of a number of distinct proteins, each capable of catalyzing the same reaction. These different enzymes, originating from the same organism or even from the same cells, have been named isozymes. They can usually be separated from each other by electrophoretic or chromatographic techniques. For example, in many tissues the oxidation of lactic acid to pyruvic acid is found to be catalyzed by five different isozymes. These isozymes differ not only in electrophoretic mobility but also in some catalytic properties. Each lactic dehydrogenase molecule is composed of four subunits, of which the two main types are the *H* and *M* subunits. The existence of five isozymes is accounted for by the fact that any combination of *H* and *M* subunits may assemble in a group of four to form an enzyme molecule. Thus there exist some lactic dehydrogenase molecules containing four *H* subunits (H_4), some with four *M* subunits (M_4), and some with intermediate properties having the compositions H_3M, H_2M_2. It may be that one of the physiological functions of isozymes is to give the cell more flexibility in metabolic regulation.

Specificity

The high degree of specificity exhibited by enzymes is one of their most characteristic properties.

Group Specificity

The majority of enzymes catalyze only one type of reaction and act on only one compound or on a group of closely related compounds. There must exist between an enzyme and its substrate a close fit, or complementarity. This relationship has been compared to that between a lock and key. In many cases, a small structural change, even in a part of the molecule remote from that altered by the enzymatic reaction, abolishes the ability of a compound to serve as a substrate. An example of an enzyme highly specific for a single substrate is urease, which catalyzes the hydrolysis of urea to carbon dioxide and ammonia. A variety of urea derivatives which have been tested are unaffected by this enzyme. On the other hand, some enzymes exhibit a less restricted specificity and act on a number of different compounds that possess a particular chemical group. This is termed group specificity.

Stereospecificity

A remarkable property of many enzymes is their high degree of stereospecificity, that is, their ability to discriminate between asymmetric molecules of the right-handed and left-handed configurations. An example of a stereospecific, as well as a group-specific, enzyme is L-amino acid oxidase. This enzyme catalyzes the oxidation of a variety of amino acids of the type $R\_CH(NH_2)COOH$. The rate of oxidation varies greatly, depending on the nature of the R group, but only amino acids of the L configuration react. Another enzyme, D-amino acid oxidase, specifically attacks amino acids of the D configuration. Conversely, when an enzyme catalyzes the formation of an asymmetric compound from symmetric substrates, the product is almost always optically active, that is, only one of the possible stereoisomers of the product is formed. For example, only the L isomer of malic acid is formed by the enzyme fumarase from fumaric acid and water, as shown in reaction (9).

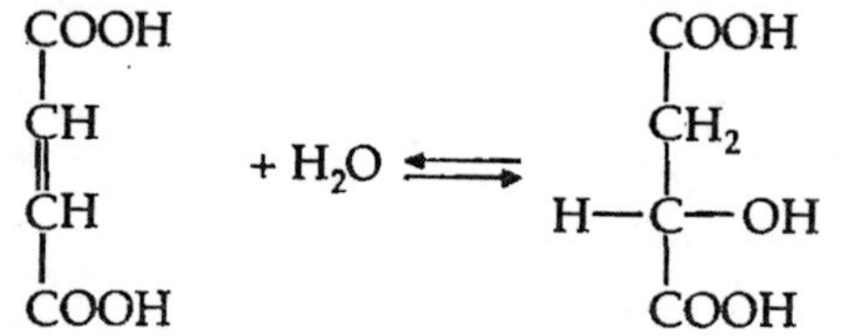

(9)

Fumaric L-Malic acid

An enzyme may also act asymmetrically on a symmetrical substrate to yield a symmetrical product. Such specificity can be detected by the use of isotopes. Thus alcohol dehydrogenase, which catalyzes the conversion of ethanol to acetaldehyde, removes a specific hydrogen atom from carbon atom 1 of ethanol. This hydrogen atom is transferred to the nicotinamide ring nicotinamide adenine dinucleotide (NAD). If reduced NAD containing deuterium (D) is used in the reversal of this reaction, asymmetrically labeled ethanol is produced, as shown in reaction (10). No deuterium is found in the acetaldehyde arising from enzymatic dehydrogenation of this ethanol.

CH_3CHO + NAD-D + H $\rightleftharpoons$

Acetaldehyde

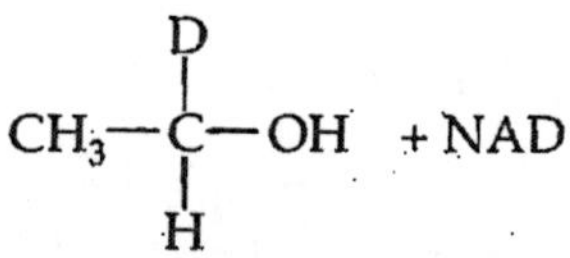

(10)

Ethanol
(asymmetrically
labeled)

The enzyme, however, cannot distinguish between hydrogen and deuterium. This is shown by the fact that when acetaldehyde containing deuterium on carbon atom 1 is hydrogenated with unlabeled NAD, the other isomer of deuterium-labeled ethanol is formed, as shown in reaction (11). On subsequent enzymatic dehydrogenation of this ethanol, all of the deuterium is found in the acetaldehyde.

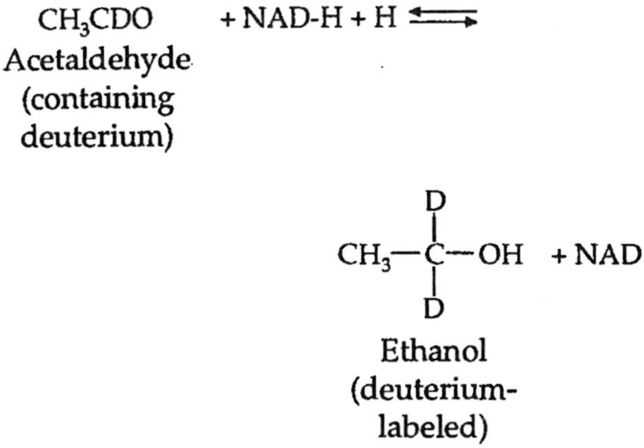

(11)

Kinetics

Much information can be obtained concerning the mechanism of an enzyme-catalyzed reaction by studying the effect of pH, temperature, and the concentration of substrates, coenzymes, inhibitors, and activators on the velocity of the reaction. Such studies were already being carried out in the nineteenth century, long before anything was known about the chemical nature of enzymes.

Michaelis-Menten Equation

It has been known for a long time that, unlike the velocity of many uncatalyzed chemical reactions, which increases linearly with the concentration of one of the reactants, the velocity of enzymatic reactions levels off and approaches a maximal value as the substrate concentration increases (Fig. 5.1). A mathematical theory accounting for this behavior was first proposed by V. Henri in 1903 and extended by L. Michaelis and L.M. Menten in 1913. The theory was based on the idea that the reaction velocity is proportional to the concentration of an enzyme-substrate complex *ES* formed by a specific combination of the enzyme *E* with its substrate *S*. This complex was assumed to react further, yielding a product *P* and regenerating a free enzyme. These reactions can be represented by reaction (12), where k_1, k_2, and k_3 are rate constants for the three reactions, whose velocities are v_1, v_2, and v_3, respectively.

$$E + S \underset{k_2}{\overset{k_1}{\rightleftharpoons}} ES \xrightarrow{k_3} E + P \qquad (12)$$

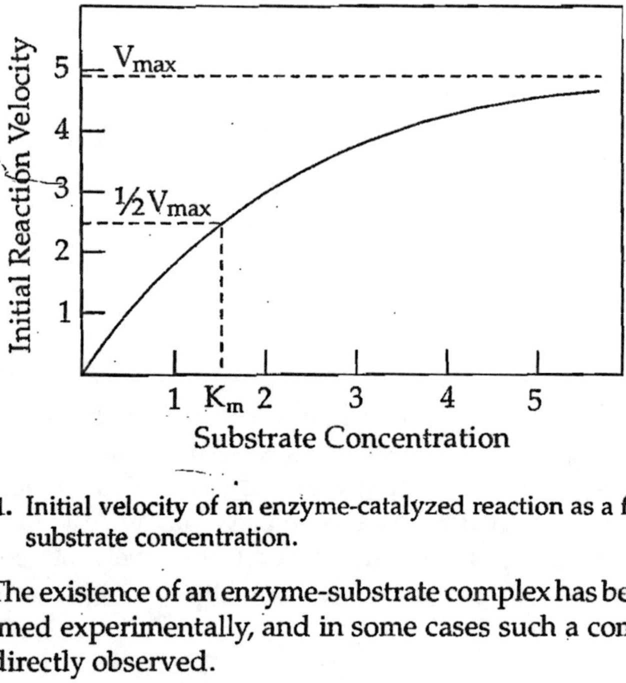

Fig. 5.1. Initial velocity of an enzyme-catalyzed reaction as a function of substrate concentration.

The existence of an enzyme-substrate complex has been amply confirmed experimentally, and in some cases such a complex has been directly observed.

Under the conditions usually employed in kinetic studies, the concentration of enzyme is very small compared to that of substrate. When enzyme and substrate are mixed, the concentration of *ES* increases rapidly until a steady state is reached in which the rate of decomposition of the complex equals its rate of formation. Such a steady state is usually reached in less than a second and may last for several minutes. During this time, the velocity of the overall reaction is essentially constant. This velocity, which may be measured as the rate of product formation, is termed the initial velocity of the reaction. An expression for the initial velocity v_0 in terms of the substrate concentration $[S]$ can now be derived simply.

According to the law of mass action, $v_1 = k_1[S][E]$, $v_2 = k_2[ES]$, and $v_3 = k_3[ES]$; by definition, $v_0 = v_3$. The total enzyme concentration

$[E_t]$ *is* $[E] + (ES)$, and the steady-state conditions state that $v_1 = v_2 + v_3$. These relations can be solved for v_0 to give Eq. (13).

$$v_0 = \frac{k_3[E_t]}{1+\dfrac{k_2+k_3}{k_1[S]}} \tag{13}$$

From Eq. (13), it is seen that, for any given substrate concentration, the initial velocity is directly proportional to the enzyme concentration. As the substrate concentration tends to infinity, the initial velocity approaches its maximal value (v_{max}), as expressed by notation (14).

$$v_{max} = k_3[E_t] \tag{14}$$

If K_m is defined as $k_2 + k_3/k_1$, the expression for v_0 becomes Eq. (15). This is the

$$v_0 = \frac{v_{max}}{1+\dfrac{k_m}{[S]}} \tag{15}$$

Michaelis-Menton equation, and the Michaelis constant K_m represents the substrate concentration required to obtain one-half the maximum velocity with a given concentration of enzyme. When k_3 is small compared to k_2, K_m approximates the dissociation constant of the enzyme-substrate complex and is therefore a measure of the affinity of an enzyme for its substrate. The equation predicts that, when v_0 is plotted as a function of [*S*], a rectangular hyperbola should be obtained, as shown in Fig. 5.1. This closely approximates the results found with a number of enzymes. Most enzymatic reactions, however, involve more than one substrate, and a number of different enzyme-substrate complexes may be formed before the appearance of the product and the regeneration of free enzyme. It may also be necessary in some cases to take into account the effect of product, inhibitors, or activators on the velocity of the reaction. With new instrumentation, it has been possible to study the kinetics of the transient, pre-steady-state portion of the reaction and to determine the value of the individual rate constants for some enzymatic reactions.

Temperature Optima

Chemical reactions are usually accelerated by an increase in temperature, because molecules require a minimum quantity of energy to react and at higher temperatures a greater proportion of the molecules possess this energy. When the rate of an enzymatic reaction is determined as a function of temperature, it is found that the rate increases with temperature up to a point and then abruptly decreases. This is because most enzymes are unstable and undergo rapid denaturation above a critical temperature. The temperature optimum for many enzymatic reactions is about 37°C (99°F).

pH Optima

Enzyme activity is also influenced considerably by the hydrogen-ion concentration of the reaction medium (Fig. 5.2). An

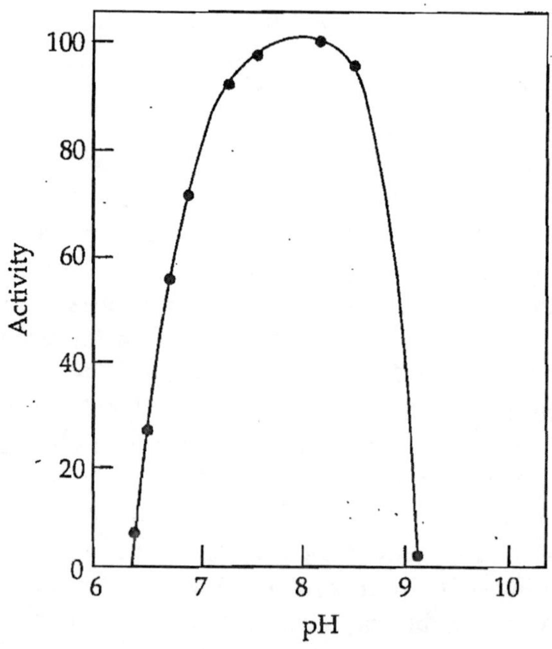

Fig. 5.2. Effect of pH on the velocity of the reaction catalyzed by the enzyme carbamyl phosphate synthetase.

enzyme may exist in several different ionic forms, and the ionic species that predominates is a function of pH. Only one of these forms can be catalytically active, or one form may be more active than the others. The pH optimum for different enzymes varies over a broad range, from about pH 2 for pepsin to about pH 10 for arginase. Some enzymes exhibit a broad optimal range extending over several pH units, while others have a very sharp pH optimum. The pH optimum for an enzyme, like the temperature optimum, may vary, depending on the substrate being used and on other experimental conditions.

Immobilization

In many instances, the use of soluble enzymes is not feasible because of high cost. Immobilization of the enzymes offers the benefit of reuse of the catalyst, thus reducing expenses, and has a second possible advantage of enhanced reaction stability.

The principal techniques for enzyme immobilization are adsorption, covalent attachment, and entrapment within polymer gels. Many enzymes have a preponderance of either positive or negative amino acid side chains on the surface, and electrostatic attraction at multiple points between a solid support and an enzyme leads to a stable complex. Hydrophobic forces may also contribute to stability of noncovalent complexes of enzyme with water-insoluble materials. Covalent attachment requires reagents that create chemical linkages (between enzyme and support). For entrapment, the enzyme is mixed with polymerizable elements, such as vinyl monomers or "prepolymers." The polymerization yields a gel containing the enzyme entrapped within it.

Properties of Immobilized Enzymes

In most cases, enzymatic activity is reduced as a result of attachment of enzymes to solid supports. Loss of activity may occur either because of chemical reaction of a functional group at or near the enzyme active center or because of physical denaturation. When charged supports are used, microenvironmental effects result in the perturbation of the pH-rate profile and the apparent reactant-binding capability of the enzyme. Both effects are reduced by increased ionic strength of the bulk solution. The transport of reactant to the surface of an enzyme particle can be limited by the unstirred (Nernst) layer surrounding the particle. The supply of reactant to enzyme molecules in the interior of the supporting particle may also be rate-limiting.

Immobilized Enzyme Reactors

The attachment of enzymes to insoluble supports permits their use in reactors which resemble those employed in the chemical process industry. Tank reactors are often preferred for intermittent, smallscale production. Viscous reactant solutions and enzyme particles with low activity might also dictate the use of tank reactors. For large-scale processes, continuous packed-bed reactors are frequently selected. Upward-flow reactors are used when the enzyme particles are compressible and when gaseous products are produced. The fluidized-bed reactor, where the enzyme particles are kept in suspension by the upward flow of reactant, can be used with viscous reactant solutions or when gaseous products are produced.

The development of ultrafiltration membranes and hollow-fiber devices has led to the construction of reactors which depend on the fact that the enzyme (of high molecular weight) can be retained by the membrane, while the product (of low molecular weight) is allowed to pass. Such reactors are not appropriate for processes in which proteases are present by choice or as contaminants; proteases are often subject to autolysis, which means that they digest themselves as well as the reactant. Contaminating proteases can severely limit the lifetime of an enzyme in this type of reactor. In contrast, when the enzyme preparation is covalently immobilized on particles, the contaminants are also immobilized (or washed through), and proteolytic degradation is eliminated.

Application of Immobilized Enzymes

A large-scale process based on immobilized enzyme technology is the production of high-fructose corn syrup, much of which is used as a sweetener for soft drinks. In this process, glucose syrup (made enzymatically from cornstarch) is passed over a bed of immobilized glucose isomerase; about 50% of the glucose is isomerized to fructose. Because fructose is sweeter than glucose, high-fructose corn syrup is sweeter than the original glucose solution; its sweetness is equivalent to a solution of sucrose (table sugar) of comparable concentration.

In the production of 6-amino penicillanic acid from penicillin G, the phenylacetyl moiety is released. from the amide linkage, but the lactam amide is left intact. The transformation occurs quantitatively under very mild conditions. This application is an excellent example of the exploitation of enzymatic specificity.

In the production of optically pure amino acids from a racemic mixture with immobilized acylase, the enzyme exhibits preference for one optical isomer; only the L enantiomer is deacylated. The resulting L-amino acid crystallizes, and the acyl-D-amino acid remains. The process can be run continuously by reracemization of the D isomer to the D,L mixture.

Chemical analysis is a very important area of application for immobilized enzymes. Here again, enzyme specificity is exploited. Enzymes can selectively transform a single compound in a complex mixture. The immobilized enzyme may be coupled with a thermal detector or an ion-selective electrode. Coating the latter with an enzyme-polymer layer results in an "enzyme electrode." An even simpler approach is to cover the tip of a conventional electrode with an ultrafiltration membrane which contains the desired enzyme solution. A cation-selective electrode immersed in a solution of glutamate dehydrogenase has been successfully used for the detection of glutamic acid.

Medical applications also appear promising. For example, a deheparinization method for blood has been proposed. The use of extracorporeal devices for blood oxygenation during bypass surgery or kidney dialysis requires heparinization of the blood to prevent clotting. However, heparin treatment can lead to hemorrhagic complications. To prevent these problems, an immobilized heparinase reactor is placed in the external circuit, prior to the point of blood reentry. Heparinase in the column completely eliminates the heparin anticoagulant activity, and heparinization is restricted to the external circuit. Immobilized heparinase may find applications in other extracorporeal devices used in enzyme therapy.

Certain types of leukemia are asparagine-dependent; when the tumor cells are deprived of asparagine, proliferation slows dramatically. The enzyme L-asparaginase degrades asparagine. Depletion of asparagine by direct injection of asparaginase into a patient is accompanied by two problems which relate to the fact that the asparaginase most commonly used is of bacterial origin. The reticuloendothelial system recognizes the enzyme as alien and clears it from the bloodstream. Also, life-threatening allergic reactions may result following administration of the foreign protein asparaginase. To immobilize the enzyme within biocompatible capsules would appear to be a potential solution to the problems. Unfortunately, immobilized enzymes are not generally good catalysts at low concentration of reactants. Since the tumor cells can survive at low levels of asparagine, effective enzyme therapy requires reduction of the asparagine concentration to very low levels. The consequence is that leukemia therapy based on the use of immobilized asparaginase remains in the experimental stage.

Production

A unique property of enzymes, both proteinaceous and RNA-derived, is the ability to catalyze chemical reactions in a mild aqueous environment, that is, with relatively low temperatures (58--212°F or 15–100°C), atmospheric pressure, and moderate pH. Thus enzymes are indispensable where harsh chemical treatment or extreme conditions are unacceptable, such as in the baking, brewing, food processing, detergent, leather, and textile industries. Novel enzymes have been developed which catalyze chemical reactions in nonaqueous organic solvents, at elevated temperatures, under conditions of extreme pH. Moreover, the chemical specificity with which enzymes catalyze chemical reactions in mild or relatively harsh conditions has led to the use of enzymes for industrial-scale biochemical processes, including synthesis of peptides, chemical feedstock compounds, and specialty compounds.

Sources

Enzymes for traditional industrial use (for example, food processing, laundry detergents, and leather processing) must be produced in large quantities and at modest cost. Consequently, they frequently are obtained as by-products from biological materials that are available in large quantities. Three common sources are plants (malt diastase, papain, bromelain, and ficin), animals (pancreatic enzymes, and digestive enzymes such as lipases, pepsin, and rennin), and microbes (amylases, proteases, cellulases, lipases, pectinases, invertase, and glucose oxidase). The number of plant- and animal-derived enzymes tends to be restricted by the availability of raw material and cost of recovery. For example, animal-derived enzymes are a low-cost by-product of the meat packing industry. However, the supply of animal-derived enzymes is restricted by the number of animals processed at meat packing plants.

To ensure a consistent supply of low-cost enzymes, industrial enzyme manufacturers have relied on microbes as a source. In fact, even specialty Industrial enzymes used in the biomedical and biotechnology industries are obtained from microbes.

The use of bacteria, fungi, and yeast can be attributed directly to the fact that they can be grown rapidly in large quantities on inexpensive culture media. Moreover, microbes can be manipulated by traditional mutagenesis and modern genetic engineering techniques to increase the yield of specific enzymes. Thus, enzymes derived from microbes are available in virtually unlimited amounts at high yield and lower cost than those derived from plants and animals.

Fermentation

The process by which microbes are cultivated is called fermentation. When used by bacterial physiologists, this term refers to the anaerobic metabolism of growth substrates. However, in industry the term is applied to any process of cultivating microbial cells in a liquid medium. Microbial fermentation takes two forms, submerged fermentation and semisolid culture. In recent times, the former has predominated, and the latter has been used to only a very limited extent. Production can be divided into three stages: ferme recovery, by which a concentrated solution of enzyme is produced, enzyme recovery, and purification.

A pure culture of the organism is inoculated into a sterilized liquid broth (250-1000 milliliters) contained in Erlenmeyer flasks.

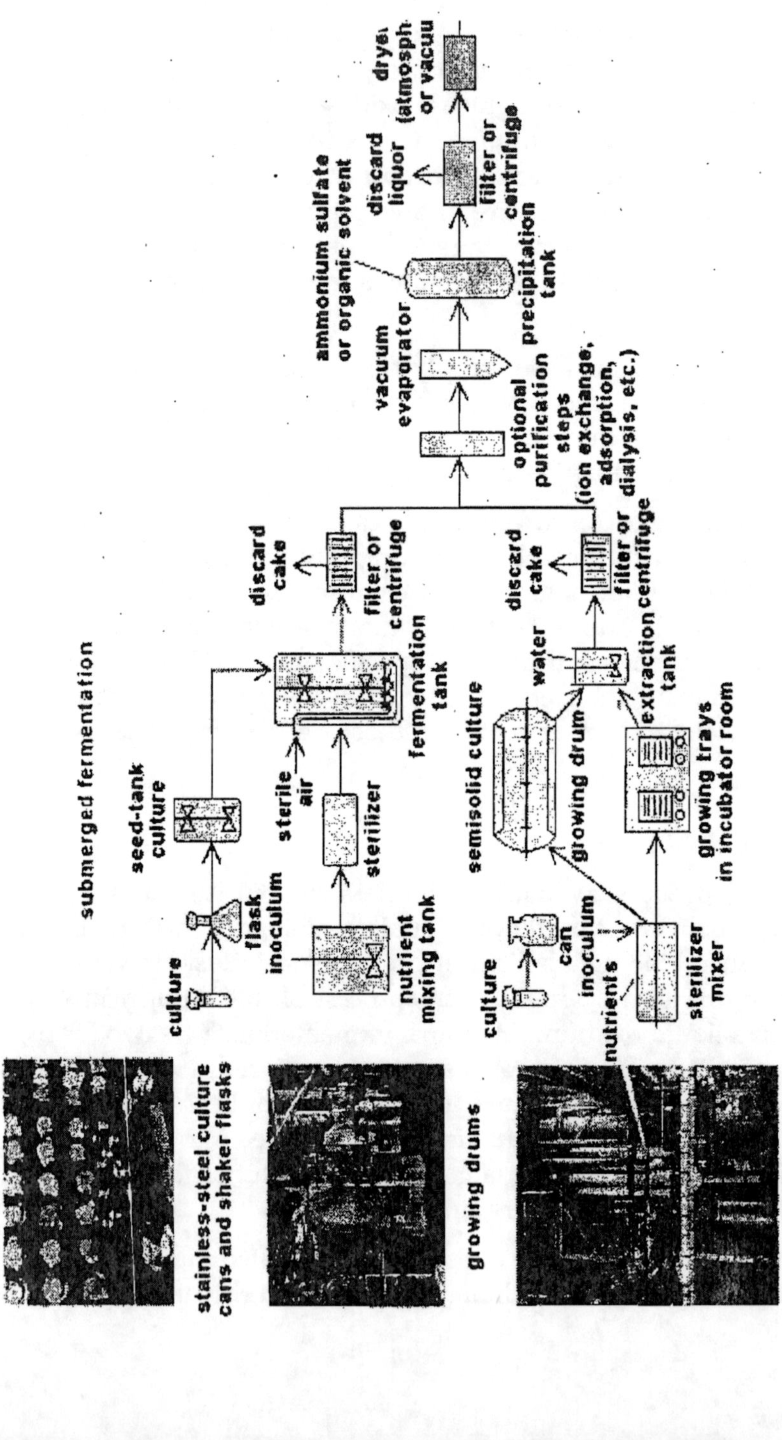

Fig. 3 Procedures for enzyme production by industrial fermentation.

The growth medium typically employed in the initial and final stages of fermentation is composed of an inexpensive carbon and energy source (potato starch, corn syrup), a nitrogen and sulfur source (hydrolyzed casein or beef extract, ammonia and sulfur salts), minerals, and growth stimulants. The growth in flasks is used to inoculate seed tanks containing 10 to several hundred liters of sterilized culture broth. In the final stage of submerged fermentation, the contents of the seed tanks are transferred to a culture medium in production tanks (4000-100,000 liters). These are stainless steel tanks equipped for the passage of air sterilized by filtration, propeller agitation, temperature control by water jacket or thermal coils, pH control, and nutrient addition (Fig. 5.3).

Depending on the microorganism, enzyme production is maximal after 1 day to a week. The fermentation tank is then emptied for enzyme recovery.

Enzyme Recovery

The first step in recovery is to concentrate the enzyme preparation. The procedure depends on whether the enzyme is contained within the cell cytoplasm (glucose oxidase, invertase) or is secreted into the medium (protease, amylase, pectinase). Enzymes contained within cells can be concentrated by centrifugation. This is followed by disintegration of cells by autolysis or a mechanical method such as high-frequency sonic vibration. In the former procedure, enzymes produced by cells degrade the microbial cell wall, resulting in cell lysis and release of the contents. Enzymes are recovered by vacuum evaporation after intact cells and other insoluble constituents have been removed by centrifugation or filtration.

Purification

Enzyme concentrates prepared as described above are frequently marketed without further purification after addition of suitable buffering and stabilizing agents. Depending on the application, further steps may be required before marketing. Additional purification may employ standard biochemical methods such as fractional precipitation by solvents (methanol, acetone, isopropanol) or salts (ammonium sulphate, calcium acetate).

Fractional precipitation is rapid, inexpensive, and applicable to very crude enzyme suspensions. It is often the first step in purification of crude broths, primarily to remove low-molecular-weight compounds with objectionable odors or tastes.

Techniques such as adsorption, ion-exchange chromatography, and gel filtration chromatography, although uncommon for enzymes used for traditional industrial applications, are typical for the purification of enzymes used in the biomedical and biotechnology industries. The added purification steps are often included to eliminate cross-contamination by enzymes or other products that would render the enzymes unsuitable for use. This is particularly true for enzymes used in the biomedical field (such as, streptokinase, ceredase, and DNase) that require intravenous administration. Enzyme preparations contaminated with endotoxin, a natural by-product of gram-negative bacteria, result in septic shock upon intravenous injection. This potentially life-threatening reaction can be eliminated only by rigorous purification of enzyme preparations prior to marketing.

Genetic Engineering

The availability of industrial enzymes has been limited by identification of suitable, naturally occurring sources of the enzyme. Genetic engineering, including recombinant DNA cloning, site-directed mutagenesis, and the polymerase chain reaction, has made it possible to extend the range of naturally occurring sources and to create completely novel enzymes that are tailored to meet the specific requirements of industrial processes.

The recombinant DNA cloning and expression of foreign genes from animal, plant, and microbial cells in high-yielding bacterial cells has generated a virtually unlimited supply of enzymes. The cloning of enzyme-encoding genes on autonomously replicating extrachromosomal DNA elements (plasmids) frequently results in the high-level expression of the cloned gene product. High-level expression results because certain plasmids exist in hundreds of copies per bacterial cell. The presence of hundreds of copies of a gene that encodes a single enzyme enables the cell to synthesize an enormous quantity of the cloned gene product compared to the source from which the gene was obtained originally. Consequently, recombinant DNA cloning of an enzyme-encoding gene can provide, in certain instances, high yields of the enzyme irrespective of the gene's source.

Recombinant DNA manipulations have been used to further enhance enzyme production from cloned genes by genetically manipulating promoters. These are DNA sequences, located upstream from the region of the DNA encoding the gene product, that are required for transcription and translation of the gene product. Enzyme manufacturers are, therefore, no longer restricted to the use of enzymes from high-yield natural sources. It is possible to identify an enzymatic activity that suits the needs of an existing process, clone the gene encoding the desired enzyme, and manipulate the gene's expression in a bacterial or other suitable host to provide a continuous and relatively low cost supply of the enzyme.

Recombinant DNA methods also provide the means to create novel enzymes. The structure of individual genes can be manipulated by fusion or by the rearrangement of distinct genes from different organisms, thereby changing the characteristics of the product encoded by the gene. In combination, these methods are used to create and modify the activities of enzymes used in industrial processes.

Another extremely powerful technology is the use of random mutagenesis of cloned genes by the polymerase chain reaction technique combined with screening of the mutagenized gene product under selective conditions. Although the approach is referred to as random mutagenesis, the mutagenesis is restricted to a specific region of a cloned gene bounded by polymerase chain reaction primers. By restricting the location of mutations, it is possible to target mutations to a general region of cloned DNA, for instance the region corresponding to the active site of an enzyme. After generating DNA that has been randomly mutated within a defined region, the DNA is cloned into appropriate sites within the gene of interest. The mutated genes are introduced into a host organism that can express the altered enzymes derived from the mutated genes. Cells containing the mutated genes are then screened under selective conditions to reveal those cells that are expressing altered gene products with novel activities. This method has been used to obtain enzymes with altered activities compatible with industrial uses. The development of enzymes with catalytic activity In nonaqueous solvents has profound implications for industrial use, particularly since many industrial processes are performed in organic rather than aqueous solvents. Moreover, combining the use of random polymerase chain reaction mutagenesis with screening under selective conditions circumvents the need for time-consuming and expensive x-ray crystallographic analysis of enzyme three-dimensional structure.

6

Bacterial Genetics

Introduction

It is the study of gene structure and function in bacteria. Genetics itself is concerned with determining the number, location, and character of the genes of an organism. The classical way to investigate genes is to mate two organisms with different genotypes and compare the observable properties (phenotypes) of the parents with those of the progeny. Bacteria do not mate (in the usual way), so there is no way of getting all the chromosomes of two different bacteria into the same cell. However, there are a number of ways in which a part of the chromosome or genome from one bacterium can be inserted into another bacterium so that the outcome can be studied.

Characteristics of Bacteria

The ways of classifying organisms are rapidly changing. Similarities in nucleotide sequences, particularly of the ribosomal ribonucleic acid (rRNA) sequences, have revealed new aspects of the evolutionary tree. All organisms have diverged from a common ancestral prokaryote whose precise location in the evolutionary tree is unclear. This has resulted in three primary kingdoms, the Archaebacteria, the Eubacteria, and the Eukaryotae. All bacteria are prokaryotes, that is, the "nucleus" or nucleoid is a single circular chromosome, without a nuclear membrane. Bacteria also lack other membrane-bounded organelles such as mitochondria or chloroplasts, but they all possess a cytoplasmic membrane. Most bacteria have a cell wall that surrounds the cytoplasmic membrane, and some bacteria also contain an outer membrane which encompasses the cell wall. Duplication occurs by a process of binary fission, in which two identical daughter cells arise from a single parent cell. Every cell in a homogeneous population of bacterial cells retains the potential for duplication. Bacteria do not possess the potential for differentiation (other than spore formation) or for forming multicellular organisms.

One of the most frequently used organisms in the study of bacterial genetics is the rod-shaped bacillus *Escherichia coli, whose normal habitat is the colon.* Conditions have been found for growing *E. coli* in the laboratory on defined media, and it is by far the best understood of all microorganisms. The single circular chromosome of *E. coli* contains about 4.5×10^6 base pairs, which is enough to make about 4500 average-size genes (1000 base pairs each). The relative positions of almost half of these genes are known. In regions where mapping studies are reasonably complete, the impression is obtained of an efficiently organized genome. Protein coding regions are located adjacent to regulatory regions. There is no evidence for significant stretches of nonfunctional deoxyribonucleic acid (DNA), and there is no evidence for introns in the coding regions. The coding regions from eukaryotic cells frequently contain introns, that is, regions that are removed by splicing the messenger RNA (mRNA) before it is translated into protein. Very little repetitive DNA exists in the *E. coli* chromosome other than the seven sequence-related rRNA genes that are dispersed at different locations on the chromosome.

Frequently, genes with a related function are tightly clustered. This clustering results from the fact that genes with a related function are commonly transcribed as a single polygenic (polycistronic) mRNA molecule. A chromosomal region transcribed in this way is called an operon. It contains a single site at one end called the promoter, which is where RNA polymerase binds just prior to the initiation of transcription.

Genetic Methodology

The first step in performing genetic research on bacteria is to select mutants that differ from wild-type cells in one or more genes. Then crosses are made between mutants and wild types, or between two different mutants, to determine dominance-recessive relationships, chromosomal location, and other properties. Various genetic methods are used to select bacterial mutants, antibiotic-resistant cells, cells with specific growth requirements, and so on.

Selection for Bacterial Mutants

Spontaneous mutation frequencies in *E. coli* vary from about 10^{-7} to 10^{-9} mutations per genome per generation. Since it is possible to plate 10^7 to 10^9 bacteria on a single standard-size growth plate, a few growth plates should contain a number of mutants of the desired type. The next consideration is to find a suitable selection procedure for detecting and isolating the desired mutants from a large, principally wild-type population. The two most popular selection techniques involve selecting for mutants with altered antibiotic resistance or nutritional requirements.

Selection for Antibiotic-Resistant Cells

Ordinary wild-type cells are sensitive to a number of bacteriostatic and bacteriocidal factors, including antibiotics, toxins produced by certain plasmid-containing cells, and bacterial viruses (bacteriophages). Mutants that are resistant to antibiotics are very useful in the analysis of biochemical pathways because the sites of action, that is, the target sites of such antibiotics, are often highly specific. For example, the target site for streptomycin action is the ribosome, as evidenced by the fact that streptomycin-resistant mutants usually have an altered ribosomal protein.

The usual procedure for isolating antibiotic-resistant mutants is simple. Approximately 10^8 cells grown in a liquid medium are spread onto a semisolid agar plate containing growth medium and the antibiotic. Only a few cells will survive and grow on the surface of the agar. These will give rise to clones, that is, a population of genetically identical cells.

Selection for Cells with Specific Growth Requirement

Wild-type *E. coli* cells that can grow on minimal medium, such as one containing only glucose ammonium chloride, and other salts, are called *prototrophs*. By contrast, auxotrophs are cells that require supplementation of the minimal medium with one or more factors.

The penicillin enrichment technique can be used to isolate an auxotroph by selecting against the prototroph. For example, a tryptophan auxotroph (*trp*⁻) will grow on glucose-minimal medium to which the amino acid tryptophan is added. To select for a *trp*⁻ strain, a large number of cells are grown on glucose-minimal medium in the presence of penicillin. Penicillin is a bacteriocidal agent that kills only growing cells by inhibiting one or more reactions in cell-wall synthesis. The *trp*⁻ cells, or for that matter any auxotrophs that cannot grow on glucose-minimal medium, will survive penicillin treatment even though they cannot grow in its presence. In fact, they survive precisely because they cannot grow. The wild-type cells that do grow in this medium will be killed by the penicillin treatment. After a suitable period, usually overnight, the cells are removed from the penicillin by centrifugation. They are resuspended in growth medium containing tryptophan and spread on an agar plate. Under these conditions the *trp*⁻ auxotrophs, which have survived the exposure to penicillin on the minimal medium, will grow.

Back mutations or revertants from an auxotrophic to a prototrophic state may be directly selected for by growth on minimal medium. Starting from a large population of auxotrophs, the only cells capable of forming clones on minimal medium would be revertants. Revertants possess either the exact reverse mutation or another mutation that compensates for the initial mutation and permits growth in the absence of added amino acid or other growth requirements.

Regulatory Gene Mutations

Certain genes that have the function of modulating the expression of other genes are known as *regulatory genes*. Mutations that affect the action of regulatory proteins are of two types: those that occur in the genes that encode the regulatory proteins, and those that affect the genetic loci where the regulatory protein interacts to modulate the level of gene expression.

In the case of tryptophan synthase, which is one of the enzymes required for tryptophan synthesis, a detailed examination has shown that the genes for tryptophan synthase are part of a multigene cluster, or operon. Expression of the operon is regulated at the promoter which is located at one end of the *trp* operon. In the *trp* operon the promoter overlaps with another site known as the operator. Whereas the promoter serves as the locus for initial binding of RNA polymerase and initiation of transcription, the operator serves as the binding site for a so-called repressor protein. When the repressor protein is bound to the operator locus, it prevents the RNA polymerase from binding to the promoter locus, thereby inhibiting the initiation of transcription and gene expression. The repressor is encoded by an unlinked gene located at a distinct site called *trpR*.

Mutations that lead to decreased or enhanced expression of the *trp* operon have been observed and analyzed in great detail. Such mutations result from changes in the DNA at the promoter-operator locus or from changes in the structure of the repressor protein, but they never affect the structure of the enzymes encoded by the operon, only their amounts. Some regulatory gene mutations cause overproduction and some cause underproduction of gene products. This is the hallmark of a mutation that influences the functioning of a regulatory protein or regulatory factor-binding site; it affects the quantity but not the quality of other gene products. Furthermore, regulatory gene mutations are frequently pleiotropic, that is, they influence the rate of synthesis of several gene products simultaneously.

Induced Mutagenesis

Frequently, geneticists want to increase the number or types of mutants that can be obtained as a result of spontaneous mutagenesis. In such instances, they treat a bacterial population with a mutagenic agent to increase the mutation frequency. This is called *induced mutagenesis*. The simplest techniques of induced mutagenesis involve measured exposure of the bacteria to a mutagenic agent, such as x-rays or chemical mutagenic agents. Such procedures have a general effect on the increase in the mutation rate.

More sophisticated procedures involve isolating the gene of interest and making a change in the desired location. This is called *site-directed mutagenesis*. The goal is usually to determine the effects of a change at a specific gene locus. The gene in question is isolated, modified, and reinserted into the organism. Discrete alterations can be made in a variety of ways on any DNA in cell-free culture, and the effect of such alterations can be subsequently tested in the organism. Most of the techniques used rely upon a site-specific restriction enzyme at some stage. The most popular procedures for directed mutagenesis are all variations of a particular procedure. This procedure begins with the limited degradation of one DNA chain of a duplex to yield a gapped molecule. The gapped DNA is annealed to a synthetic oligonucleotide containing discrete differences from the original DNA. After completing the ligation process, the resulting heteroduplex can be reinserted into the organism, and by suitable selection procedures mutants can be isolated that contain specific changes dictated by the sequence in the synthetically produced oligonucleotide.

Gene Mapping

Bacteria do not mate to form true zygotes, but they are able to exchange genetic information by a variety of processes in which partial zygotes (merozygotes) are formed. The first type of genetic exchange between bacteria to be observed was transformation. Naturally occurring transformation involves the uptake of DNA. This phenomenon is observed only for a limited number of bacterial species and is a relatively difficult technique to use for gene manipulation. For these reasons and because more powerful approaches to gene mapping were discovered soon afterward, transformation has received relatively narrow use as a means for gene mapping. In 1946 direct chromosomal exchange by conjugation between *E. coli* cells was discovered by J. Lederberg and E. Tatum, and in 1951 transduction, the virus-mediated transfer of bacterial genes, was discovered. Both conjugation and transduction provide facile, generally applicable methods for moving part of the bacterial chromosome from one cell to another. The discovery of bacterial transposons in the 1970s has been useful in marking and mobilizing genes of interest. The purely genetic approaches to mapping have been supplemented by the biochemical approaches of hybrid plasmid construction and DNA sequence analysis. A detailed genetic and physical map of the entire bacterial chromosome is within grasp.

Conjugation

The naturally occurring exchange of genetic information in *E. coli* bacteria takes place between cells of different mating types.

Donor cells are defined as bacteria capable of transferring genes, and recipient cells are defined as bacteria capable of accepting and integrating such genes. The difference between a donor and a recipient cell is due to a plasmid known as *F factor*. The F factor is a circular DNA molecule with a molecular weight of about one-fortieth that of the bacterial host chromosome. Donor cells (F^+ cells) arise from bacteria that contain the F factor. Within an F^+ population, cells that transfer bacterial genes represent distinct variants. These variants, called *Hfr's*, have integrated the F factor into the host chromosome, and the site of integration determines the order of transfer of bacterial genes when an Hfr is mated with an F^- cell. The order of transfer of markers may be determined by interrupting the mating process at different times after mixing the two cell types. The so-called interrupted mating procedure has been carried out on many different Hfr's that have integrated the F factor at different locations and in either location. The results from such conjugations have permitted a rough mapping of the bacterial chromosome. Indeed, it was studies of this sort that led to the prediction that the bacterial chromosome was circular, a prediction that was verified by physical measurements many years later.

On rare occasions, the integrated F factor in the Hfr can reversibly excise to give back the original F^+ cell. On even rarer occasions, the F factor excises imperfectly, taking with it a part of the bacterial chromosome. Such F′ factors have been used as a valuable source of discrete segments of the bacterial chromosome; they can readily be transferred to F″ cells. They have been most valuable in determining dominance-recessive relationships for alleles of the same genes.

Transduction

Transfer of bacterial segments can be done by seduction, that is, the transfer of F′ particles, but a far more convenient procedure for most applications involves transport by bacteriophages. Transfer of bacterial DNA by bacteriophages is called *transduction*. Transduction has provided the most valuable technique for bacterial gene mapping and for the isolation of discrete segments of the bacterial chromosome. A number of bacteriophages are used in transduction. They have different properties, and each one has particular advantages for gene manipulation. Some bacteriophages, such as the l bacteriophage, transfer only specialized regions of the bacterial genome. For this reason, l is referred to as a specialized transducing virus. Other bacteriophages, such as P1, which can transfer virtually any part of the bacterial genome are referred to as generalized transducing viruses. To obtain a P1 preparation that carries parts of a particular bacterial chromosome, the bacterial strain is infected with P1. The infection results in the production of a large number of normal P1 virus particles and a small mixed population of P1 transducing particles. The P1 transducing particles carry segments of the bacterial genome instead of the viral genome. When they infect a bacterial cell, the segment of bacterial genome enters the cell.

For a high-resolution mapping of bacterial genes, generalized transduction is a virtually indispensable tool. Bacterial genes that are transduced into *E. coli* by P1 transducing particles cannot replicate or survive unless they become integrated into the host genome. In so doing, they replace the homologous segment in the bacterial genome. Integration requires a minimum of two crossovers. In three-factor crosses, the middle marker may be identified by the least frequent recombinant class. Exclusive exchange of the middle marker requires two crossovers also. It occurs with a lower than average probability because the crossover points are restricted so that only the middle marker exchanges.

Transposons

Chromosomes might be thought of as being composed of two types of genetic elements: those that maintain a fixed location in a larger stable structure, and those that are designed to move from one location to another. The latter type is known by the general term mobile genetic element. Transposons are a class of mobile genetic elements that are commonly found in bacterial populations. Most transposons have three characteristic properties: they carry readily selectable marker genes; they insert at many different locations in the bacterial chromosome; and they are mutagenic in regions adjacent to the site of insertion.

Expression of Bacterial Genes

At any given time, only a small percentage of the *E. coli* genome is being actively transcribed. The remainder of the genome is either silent or being transcribed at a very low rate. When growth conditions change, some active genes are turned off and other, inactive genes are turned on. The cell always retains its totipotency, so that within a short time (seconds to minutes), and given appropriate circumstances, any gene can be fully turned on. The maximal activity for transcription varies from gene to gene. For example, a b-galactosidase gene makes about one copy per minute, and a fully turned-on biotin synthase gene makes about one copy per 10 min. In the maximally repressed state, both of these genes express less than one transcript per 10 min.

The level of transcription for any particular gene usually results from a complex series of control elements organized into a hierarchy that coordinates all the metabolic activities of the cell. For example, when the rRNA genes are highly active, so are the genes for ribosomal proteins, and the latter are regulated in such a way that stoichiometric amounts of most of the ribosomal proteins are produced. When glucose is abundant, most genes involved in processing more complex carbon sources are turned off in a process called *catabolite repression*. If the glucose supply is depleted and lactose is present, the genes involved in lactose breakdown (catabolism) are expressed. In *E.* coli the production of most RNAs and proteins is regulated exclusively at the transcriptional level, although there are notable exceptions. Rapid response to changing conditions is ensured partly by a short mRNA lifetime, on the order of 1 to 3 min for most mRNAs. Some mRNAs have appreciably longer lifetimes (10 min or more), and the consequent potential for much higher levels of protein synthesis per mRNA molecule. These atypical mRNAs, at least in some instances, also may be subject to translational control. Finally, the fine-level control for any particular enzyme system is subject to regulation by activators or inhibitors directly modulating the enzyme activities.

Rel Gene

The ribosome is the site of protein synthesis. Bacterial ribosomes are composed of 3 rRNAs and about 50 ribosomal proteins. *Escherichia coli* cells, under conditions of rapid growth, contain about 10,000 ribosomes. There are seven operons located at distinct and separate points on the bacterial chromosome which make similar rRNA transcripts. The transcription of the rRNA operons is controlled by the compound guanosine tetraphosphate, ppGpp, known as *magic spot*. This compound has pyrophosphates at the 3′ and the 5′ carbons of the ribose sugar. Under conditions of amino acid deprivation, the level of magic spot rises. It combines with the RNA polymerase, changing the structure of the RNA polymerase so that the latter has a low affinity for the promoters of the rRNA genes. The result is that less rRNA is made. The ppGpp compound has a similar effect on the expression of genes for the rRNA proteins. The mRNAs for the ribosomal protein genes are also controlled at the translation level by a feedback inhibition mechanism. These proteins normally combine with the rRNA to make ribosomes. However, if there is an excess of ribosomal proteins over the available rRNA, the ribosomal protein will combine with its own message so as to prevent further translation of the message. In this way the amount of ribosomal protein never reaches large excess but always is synthesized in slight excess over the available level of rRNA.

Gene Regulation for DNA Repair

The cell possesses an intricate system (the *din* genes) for regulating the synthesis of genes involved in DNA repair. The *din* genes are under the negative control of the lex repressor. Unlike most repressors, including the lac repressor, the lex repressor action is not reversibly controlled by complexing with a small molecule. Rather, it is broken down by a highly specific protease. This protease is carried by the recA protein and is not active under normal growth conditions. However, when the growing cells are exposed to ultraviolet light or other agents that damage DNA, the recA protease becomes active. It is believed that this is due to a complex formed between the recA protease and a DNA fragment. Once the protease becomes active, it breaks down all of the IexA protein in the cell. The genes repressed by lexA then become active and stay active until the DNA damage is repaired. After the DNA damage has been repaired, the recA protease returns to its normal inactive state, and the concentration of lexA repressor reaches a level where it silences the *din* genes and everything returns to normal.

7

Human-Factors Engineering

Introduction

Human factor engineering is also called *ergonomics, cr human engineering,* science dealing with the application of information on physical and psychological characteristics to the design of devices and systems for human use.

The term human-factors engineering is used to designate equally a body of knowledge, a process, and a profession. As a body of knowledge, human-factors engineering is a collection of data and principles about human characteristics, capabilities, and limitations in relation to machines, jobs, and environments. As a process, it refers to the design of machines, machine systems, work methods, and environments to take into account the safety, comfort, and productiveness of human users and operators. As a profession, human-factors engineering includes a range of scientists and engineers from several disciplines that are concerned with individuals and small groups at work.

The terms human-factors engineering and human engineering are used interchangeably on the North American continent. In Europe, Japan, and most of the rest of the world the prevalent term is ergonomics, a word made up of the Greek words, *ergon,* meaning "work," and *nomos*, meaning "law." Despite minor differences in emphasis, the terms human-factors engineering and ergonomics may be considered synonymous. Human factors and human engineering were used in the 1920s and '30s to refer to problems of human relations in industry, an older connotation that has gradually dropped out of use. Some small specialized groups prefer such labels as bioastronautics, biodynamics, bioengineering, and mannedsystems technology; these represent special emphases whose differences are much smaller than the similarities in their aims and goals.

The data and principles of human-factors engineering are concerned with human performance, behavior, and training in man-machine systems; the design and development of man-machine systems; and systems-related biological or medical research. Because of its broad scope, human-factors engineering draws upon parts of such social or physiological sciences as anatomy, anthropometry, applied physiology, environmental medicine, psychology, sociology, and toxicology, as well as parts of engineering, industrial design, and operations research.

The Human-Factors Approach to Design

Two general premises characterize the approach of the human-factors engineer in practical design work. The first is that the engineer must solve the problems of integrating humans into machine systems by rigorous scientific methods and not rely on logic, intuition, or common sense. In the past the typical engineer tended either to ignore the complex and unpredictable nature of human behavior or to deal with it summarily with educated guesses. Human-factors engineers have tried to show that with appropriate techniques it is possible to identify man-machine mismatches and that it is usually possible to find workable solutions to these mismatches through the use of methods developed in the behavioral sciences.

The second important premise of the human-factors approach is that, typically, design decisions cannot be made without a great deal of trial and error. There are only a few thousand human-factors engineers out of the thousands of thousands of engineers in the world who are designing novel machines, machine systems, and environments much faster than behavioral scientists can accumulate data on how humans will respond to them. More problems, therefore, are created than there are ready answers for them, and the human-factors specialist is almost invariably forced to resort to trying things out with various degrees of rigour to find solutions. Thus, while human-factors engineering aims at substituting scientific method for guesswork, its specific techniques are usually empirical rather than theoretical.

The Man-Machine Model

Human-factors engineers regard humans as an element in systems, and a man-machine model is the usual way of representing that relationship. The simplest model of a man-machine unit consists of an individual operator working with a single machine. In any machine system, the human operator first has to sense what is referred to as a machine display, a signal that tells him something about the condition or the functioning of the machine. A display may be the position of a pointer on a dial, a light flashing on a control panel, the readout of a digital computer, the sound of a warning buzzer, or a spoken command issuing from a loudspeaker.

Having sensed the display, the operator interprets it, perhaps performs some computation, and reaches a decision. In so doing, the worker may use a number of human abilities, including the ability to remember and to compare current perceptions with past experiences, to coordinate those perceptions with strategies formed in the past, and to extrapolate from perceptions and past experiences to solve novel problems. Psychologists commonly refer to these activities as higher mental functions; human-factors engineers generally refer to them as information processing.

Having reached a decision, the human operator normally takes some action. This action is usually exercised on some kind of a control--a pushbutton, lever, crank, pedal, switch, or handle. The action upon one or more of these controls exerts an influence on the machine and on its output, which in turn changes the display, so that the cycle is continuously repeated.

A man-machine system does not exist in isolation; it exists in an environment of some sort. Since the nature of this environment influences the operator's efficiency and performance, the human-factors engineer must be concerned with such environmental factors as temperature, humidity, noise, illumination, acceleration, vibration, and noxious gases and contaminants.

A Man-Machine Example

Driving an automobile is a familiar example of a simple man-machine system. In driving, the operator receives inputs from outside the vehicle (sounds and visual cues from traffic, obstructions, and signals) and from displays inside the vehicle (such as the speedometer, fuel indicator, and temperature gauge). The driver continually evaluates this information, decides on courses of action, and translates those decisions into actions upon the vehicle's controls--principally the accelerator, steering wheel, and brake. Finally, the driver is influenced by such environmental factors as noise, fumes, and temperature.

The simple man-machine model provides a convenient way for organizing some of the major concerns of human engineering: the selection and design of machine displays and controls; the layout and design of workplaces; design for maintainability; and the work environment.

Human Factors in Large Systems

No matter how important it may be to match an individual operator to a machine, some of the most challenging and complex human problems arise in the design of large man-machine systems and in the integration of human operators into these systems. Examples of such large systems are a modern jet airliner, an automated post office, an industrial plant, a nuclear submarine, and a space vehicle launch and recovery system. In the design of such systems, human-factors engineers study, in addition to all the considerations previously mentioned, three factors: personnel, training, and operating procedures.

Systems are generally designed for particular kinds of operators. A space vehicle, for example, is designed for a highly select handful of astronauts. Automobiles, on the other hand, are designed to accommodate a wide spectrum of people. In large systems, the specification of personnel requirements is an integral part of systems design.

Personnel are trained; that is, they are given appropriate information and skills required to operate and maintain the system. System design includes the development of training techniques and programs and often extends to the design of training devices and training aids.

Instructions, operating procedures, and rules set forth the

duties of each operator in a system and specify how the system is to function. Tailoring operating rules to the requirements of the system and the people in it contributes greatly to safe, orderly, and efficient operations.

Applications of Human-Factors Engineering

The basis of human-factors engineering--the consideration of information about human users in the design of tools, machines, jobs, and work environments--has always been present. One of the oldest and most efficient of human implements, the scythe, shows a remarkable degree of human-factors engineering, undoubtedly reflecting modifications made over many centuries: the adroitly curved handle and blade and the peg grasp for the left hand. All of this is in sharp contrast with the conventional snow shovel, a modern implement of generally poor design that has been blamed for many a wintertime back strain.

The need for a more formal approach to these human problems was created when machines became vastly more complex than they had ever been. High-speed jet aircraft, computers, radar, nuclear submarines, communication satellites, space vehicles--all these are products of the past few decades. The fantastic growth in the number and complexity of machines has created entirely new problems about the use of human operators and the way they can be integrated into systems. Moreover, the solution to these new problems cannot be found in the collective wisdom of society. For example, not long ago no one had any way of predicting with any certainty how astronauts could or would function in a weightless environment. Human-factors engineering is, therefore, a child of the times, born of a mechanized civilization.

Applications of human-factors engineering have been made to such simple devices as highway signs, telephone sets, hand tools, stoves, and to a host of modern, sophisticated complexes such as data processing systems, automated factories and warehouses, robots, and space vehicles.

The experience gained in devising these systems has contributed to the realization that even relatively simple devices raise unexpectedly important questions on human use--questions that conventional engineering practice frequently cannot answer.

Push-Button Telephone

The modern push-button telephone handset provides a good example of a relatively simple device that has required a great deal of human-factors engineering. The layout of the keys in the four rows of three buttons, for example, was selected only after extensive tests on a variety of arrangements: circular, two vertical rows of five buttons, two horizontal rows of five, and a diagonal pattern; the arrangement of the numerals and letters on the keys, in the order of left to right and from top to bottom, was chosen as superior to other arrangements such as that used on many desk calculators, in which the numbers increase from bottom to top. The top-to-bottom design decision was not simply a matter of logic; tests showed that people actually made fewer errors and took less time with that arrangement than they did with the calculator arrangement. Other human-factor considerations in the design of the push-button keyset were the size and style of numerals and letters for maximum legibility, the optimum sizes and spacing of the keys, and the proper force-displacement characteristics of the keys to provide tactile feedback or "feel" when the buttons are depressed.

Similar factors were considered in designing the shape of the handset itself. The locations, separations, and angles between the earpiece and mouthpiece were determined so that the assembly would fit comfortably around the greatest number of different human faces; and the weight of the handset was designed to be neither too light nor too heavy. In recent years the careful, "userfriendly" design of conventional telephone sets has become more apparent in contrast to some of the new arrivals in the telephone marketplace, which are generally inferior in design and quality.

Space Suit

The designing of a much more complicated device, such as a space suit, presents more intricate problems. A space suit is a complete miniature world, a self-contained environment that must supply everything needed for an astronaut's life, as well as comfort. The suit must provide a pressurized interior, without which an astronaut's blood would boil in the vacuum of space. The consequent pressure differential between the inside and the outside of the suit is so great that when inflated the suit becomes a distended, rigid, and unyielding capsule. Special joints were designed to give the astronaut as much free movement as possible. The best engineering has not been able to provide as much flexibility of movement as is desirable; to compensate for that lack, attention has been directed toward the human-factors design of the tools and devices that an astronaut must use.

In addition to overcoming pressurization and movement problems, a space suit must provide oxygen; a system for removing excess products of respiration, carbon dioxide and water vapour; protection against extreme heat, cold, and radiation; protection for the eyes in an environment in which there is no atmosphere to absorb the sun's rays; facilities for speech communication; and facilities for the temporary storage of body wastes. This is such an imposing list of human requirements that an entire technology has been developed to deal with them and, indeed, with the provision of simulated environments and procedures for testing and evaluating space suits.

Typewriter Keyboard

Not all human-factors engineering and design is commercially successful. An example is the typewriter keyboard. Several alternative layouts, which are demonstrably superior from a human-factors point of view, have been proposed, beginning as far back as the 1920s. Despite test results which show that alternative layouts are easier to learn, create less operator fatigue, and permit faster typing, the traditional layout persists and now has been carried over into the design of millions of personal computers. In this case, inertia and resistance to change have been more formidable obstacles to efficient ergonomic design than the design itself.

Social Problems

The telephone, the space suit, and the typewriter keyboard are but three out of thousands of examples that might have been selected to show how human-factors engineering has been consciously applied to solve technological problems. The same human-factors principles and methods have also been applied to a variety of social problems, such as individualized computer-assisted instruction, nonlethal antiriot equipment for law enforcement agencies, antiterrorist architecture for public buildings, and people movers for airport and urban transportation departments. The modern concern with man's relationship to the total environment implies a much-broadened definition of human-factors engineering and an increasing supply of problems for ergonomic engineers in the future.

8

Antibody

Introduction

It is a protein found principally in blood serum and characterized by a specific reactivity with the corresponding antigen. Antibodies are important in resistance against disease, in allergy, and in blood transfusions, and can be utilized in laboratory tests for the detection of antigens or the estimation of the immune status.

Induction

Antibodies are normally absent at birth unless derived passively from the mother through the placenta or colostrum. In time, certain antibodies appear in response to environmental antigens, although the inducers have not been determined for all normal antibodies, particularly the human blood-group isoagglutinins. Antibodies are also induced by artificial immunization with vaccines or following natural infections. The resulting antibody level declines over a period of months, but rapidly increases following renewed contact with specific antigen, even after a lapse of years. This is known as an *anamnestic* or *booster response*.

Antibody reactivity results in precipitation of soluble antigens, agglutination of particulate antigens, increased phagocytosis of bacteria, neutralization of toxins, and dissolution of bacterial or other cells specifically sensitive to their action; the antibodies so revealed are termed precipitins, agglutinins, opsonins, antitoxins, and lysins. One antibody may give many such reactions, depending on conditions, so these classifications are not unique or exclusive.

Groups

Three principal groups (IgG, IgM, IgA) and two minor groups (IgD, IgE) of antibodies are recognized. These all form part of the wider classification of immunoglobulins. The basic molecule of each class of immunoglobulin comprises four polypeptide chains - two identical heavy chains and two identical light chains - which are held together by disulfide bonds. The molecular weight of the heavy chains is about 50,000, that of the light chains about 20,000. Both heavy and light chains consist of a variable region and a constant region. The variable region contains large invariant stretches which are interspersed with "hot spots" of variability (hypervariability). Antigen-binding activity resides in the variable region. The monomeric antibody molecule has two combining sites, each composed of parts of the hypervariable regions of both a heavy chain and a light chain. Antibody diversity is generated by amino acid substitutions that occur within the hypervariable regions and result in unique antigen-binding structures.

Immune Response

During the 1970s and 1980s, considerable advances were made in understanding the immune response. The structure of the immunoglobulin molecule was defined; second, it was determined that the immunoglobulin molecule was composed of a constant region and a variable region, suggesting that there could be a structural basis for antibody diversity. While the molecular basis of specificity was being elaborated, cellular immunologists established the validity of the clonal selection theory. This theory states that a lymphocyte synthesizes antibodies of only one specificity and is preprogrammed for this specificity. Therefore, the events responsible for the synthesis of antibodies of a certain specificity have occurred before the introduction of antigen. From molecular biology, there are indications that two seperate genes code for a single chain of an antibody molecule.

Immunoglobulin Production

The introduction of antigen into an animal causes a complex

series of events affecting the *B* and *T* lymphocytes; *B* cells are the effector cells, and certain *T* cells serve as helper cells. Immunoglobulin molecules found on the surface (sIg) of *B* cells act as receptors for antigen. Specificity of this immunoglobulin is the same as the antigen specificity of the immunoglobulin secreted by the cell. As *B* cells differentiate, the first immunoglobulin expressed on their surface is IgM (sIgM); as *B* cells mature, they express another isotype of immunoglobulin on their surface, that is, sIgD. The sIgM and sIgD both have the same specificity and idiotype. The membrane form of the immunoglobulin molecules is slightly larger than the secreted form because of the presence of a hydrophobic region which anchors them to the membrane. Reorganization of the immunoglobulin molecules in the cell membrane by patching and capping is important for signal transduction from the surface of the cell to the nucleus. The *T* cells do not have surface immunoglobulin. Activation of *B* cells requires that they interact at the same time with helper *T* cells or bind certain soluble growth and differentiation factors. Activation of *B* cells has two phases, proliferation and differentiation: proliferation expands the number of cells capable of reacting with antigenic substances that have been introduced into the individual, and differentiation transforms the *B* cells into antibody – secreting cells.

Monoclonal Antibodies

Because most antigens contain several antigenic determinants, when an antigen is introduced into an animal, a seperate clone of antibody-producing cells is generated for each of the antigenic determinants. The individual clones can be seperated and cultured, making possible the production of antibody against one specific determinant. The principle of production of these monoclonal antibodies is that normal antibody-forming cells (*B* cells) cannot be grown and perpetuated in tissue culture. However, tumors of *B* cells (myelomas) can be grown indefinitely in culture. Therefore, an antibody-forming cell must fuse with a myeloma cell to form a hybrid cell, termed a hybridoma. Single hybridoma cellsare distributed into tissue culture microwells and grown. The supernatants are tested for antibody of the desired specificity. Cells from wells that contain the appropriate antibody are further cloned. This method allows for the production of large amounts of monospecific antibody.

The development of the technology for producing monoclonal antibodies revolutionized the uses of antibodies in all of biology and medicine. It is now possible to produce antibodies for almost any position on a protein. Antibodies and enzymes are both proteins whose function depends on their ability to bind specific target molecules. Enzymatic reactions differ from immunological reactions in that the binding of substrate to enzyme leads to chemical catalysis. Enzymes catalyze the hydrolysis of proteins by combining with the protein in a transition state of the reaction. Thus, instead of a monoclonal antibody binding directly to the virus or tumor antigen (a reaction which requires the participation of some other effector system such as the complement system to rid the body of the target), such an antibody could catalyze specific protein cleavages which would directly inactivate viruses or kill cells without the need of additional effector system activation.

Unfortunately, there is a major problem with the use of monoclonal antibodies in living organisms. Almost all monoclonal antibodies originate in mice, and the murine immunoglobulin serves as an antigen, frequently acting immunogenic in human recipients.

9

Vaccination

Introduction

It is an active immunization against a variety of microorganisms or their components, with the ultimate goal of protecting the host against subsequent challenge by the naturally occurring infectious agent.

The terms vaccine and vaccination were originally used only in connection with Edward Jenner's method for preventing smallpox, introduced in 1796. In 1881 Louis Pasteur proposed that these terms should be used to describe any prophylactic immunization. Vaccination now refers to active immunization against a variety of bacteria, viruses, and parasites (for example, malaria and trypanosomes).

During the last 200 years, vaccination has controlled nine major diseases: smallpox, diphtheria, tetanus, yellow fever, pertussis, poliomyelitis, measles, mumps, and rubella. In the United States and other developed countries, where programs have been expanded to immunize virtually all children under 2 years of age, most vaccine-preventable diseases have been reduced by 95% or more from prevaccine highs (Table 9.1). In the case of smallpox, the dream of eradication has been fulfilled. Vaccinations against influenza, hepatitis *B*, pneumococci, and hemophilus influenza type *b* have made major headway against these infections.

Implicit within Jenner's method of vaccinating against smallpox was the recognition of immunologic cross-reactivity together with the notion that protection can be obtained through active immunization with a different, but related, live virus. It was not until the 1880s that the next immunizing agents, vaccines against rabies and anthrax, were introduced by Pasteur. Two facts of his experiments on rabies vaccines are particularly noteworthy.

Table 9.1. Past maximum and current morbidity from vaccine preventable diseases

Disease	*Number of cases Maximum (year)*	*In 1998 (provisional)*
Diphtheria	206,939 (1921)	1
Measles	894,134 (1941)	100
Mumps	152,209 (1968)	666
Pertussis	265,269 (1934)	7405
Polio (paralytic)	21,269 (1952)	1
Rubella	57,686 (1969)	364
Congenital rubella syndrome	20,000 (1964-5)	7
Tetanus*	1,560 (1923)	41

*Deaths (death registration states only) in 1923; cases in 1998.

First, Pasteur found that serial passage of the rabies agent in rabbits resulted in a weakening of its virulence in dogs. For the next 100 years, Pasteur's empirical approach for attenuating the virulence of a live virus by repeated passages in cells of species different from the natural host remained the principal empirical method for developing attenuated-live-virus vaccines. During multiple passages in an animal or in tissue culture cells, mutations accumulate as the virus adapts to its new environment. These mutations adversely affect virus reproduction in the natural host, resulting in lessened virulence. Only as the molecular basis for virulence has begun to be elucidated by modern biologists has it become possible to deliberately remove the genes promoting virulence so as to produce attenuated viruses.

Second, Pasteur demonstrated that rabies virus retained immunogenicity even after its infectivity was inactivated by formalin and other chemicals, thereby providing the paradigm for one class of noninfectious virus vaccine, the "killed"-virus vaccine.

Attenuated-live and inactivated vaccines are the two broad classifications for vaccines. Anti-idiotype antibody vaccines and deoxyribonucleic acid (DNA) vaccines represent innovations in inactivated vaccines. Recombinant-hybrid viruses are novel members of the live-virus vaccine class recently produced by genetic engineering.

Attenuated-Live Vaccines

Because attenuated-live-virus vaccines reproduce in the recipient, they provoke both a broader and more intense range of antibodies and T-lymphocyte-associated immune responses than noninfectious vaccines. Live-virus vaccines have been administered subdermally (vaccinia), subcutaneously (measles), intramuscularly (pseudorabies virus), intranasally (infectious bovine rhinotracheitis), orally (trivalent Sabin poliovirus), or by oropharyngeal aerosols (influenza). Combinations of vaccines have also been used. When combined live measles, rubella, and mumps vaccines have been given by injection, the antibody response to each component has been comparable with the antibody response to the individual vaccines given separately.

Many organisms produce their pathophysiologic effects through acute infections localized to the gastrointestinal, pulmonary, nasopharyngeal, and genitourinary surfaces. These areas are bathed in mucus that contains highly important secretory immunoglobulin A (IgA). Subcutaneous, intravenous, and intramuscular immunization regimens, though effective in inducing IgG-type antibodies, are almost universally ineffective at the induction of secretory IgA antibodies. In contrast, oral administration is a feasible alternative to stimulate the common mucosal immune system, as with the live-attenuated polio vaccine. Live-virus vaccines administered through a natural route of infection often induce local immunity, which is a decided advantage. However, in the past, attenuated-live virus vaccines have been associated with several problems, such as reversion to virulence, natural spread to contacts, contaminating viruses, lability, and viral interference.

Noninfectious Vaccines

Noninfectious vaccines include inactivated killed vaccines, subunit vaccines, synthetic peptide and biosynthetic polypeptide vaccines, oral transgenic plant vaccines, anti-idiotype antibody vaccines, DNA vaccines, and polysaccharide-protein conjugate vaccines. With most noninfectious vaccines a suitable formulation is essential to provide the optimal antigen delivery for maximal stimulation of protective immune responses. Development of new adjuvant and vector systems is pivotal to produce practical molecular vaccines.

Killed Vaccines

The inactivated forms of the toxins produced by the tetanus and diphtheria bacteria, called *toxoids*, were among the first vaccines developed. For human use, toxoid vaccines must be highly purified to remove extraneous materials. Killed-virus vaccines consist of partially purified virus particles whose infectivity has been destroyed by treatment with chemicals or with radiation. These virus particles retain the ability to elicit an immune response and to protect from clinical disease. Though generally safer and more stable than attenuated-live-virus vaccines, killed-virus vaccines require multiple doses of high concentrations of antigen sometimes administered with adjuvant, that is, a substance that enhances the potency of the antigen. The induction of cellular immune responses may be poor and the protection induced by killed-virus vaccines may be of short duration, so that booster vaccinations are needed. Vaccines produced by killing viruses have been known to contain surviving, infectious particles that have caused outbreaks of disease (Cutter polio vaccine, foot-and-mouth-disease vaccine). Although incapable of producing direct cellular destruction, noninfectious vaccines may elicit abnormal hypersensitivity reactions.

Through the use of the oral, attenuated-live (Sabin) vaccine and the inactivated (Salk) vaccine, paralytic polio has been virtually eliminated from the United States. The last naturally occurring case of paralytic polio in the United States was in 1979. In 1997 and 1998, there were only six reported cases. All were associated with receipt of the trivalent oral, attenuated-live vaccine. An all-inactivated polio vaccine schedule has been recommended for routine childhood vaccination beginning January 1, 2000.

In the 1970s, concerns developed that whole-cell pertussis vaccines caused permanent brain damage in some children. This led to a sharp decline in pertussis vaccination, waning immunity, and outbreaks of disease in several countries. Even though the general risk of serious postvaccination neurologic damage was not confirmed, a strong feeling persisted that safer and more efficacious pertussis vaccines were needed. Acellular pertussis vaccines consisting of pertussis toxoid (and in some cases, highly purified filamentous hemagglutinin and pertactin proteins) were recommended in the United Kingdom, Sweden, and Japan, and were approved in 1996 in the United States.

Polysaccharide Vaccines

Polysaccharide vaccines consist of the polysaccharide coats, or capsules, of encapsulated bacteria. Although polysaccharides are recognized mainly by T-cell-independent mechanisms, they do not effectively produce high-level, high-affinity antibodies, nor do they induce the T-cell memory required for booster responses. However, the immunogenicity of polysaccharides can be significantly increased by covalently linking them to carrier proteins. This approach is thought to work through the recruitment of T-cell help. For example, Hemophilus influenza b vaccines developed with this conjugate approach have been very successful in inducing high, boostable, protective antibody levels in infants and have dramatically reduced Hemophilus influenza b-associated disease in countries where they are in general use.

Subunit Vaccines

Subunit vaccines consist of immunogenic viral proteins stripped free from whole virus particles, then purified from other irrelevant components, thereby reducing the risk of adverse reactions and residual infectious virus. Subunit vaccines require the addition of adjuvants or their formation into forms such as lipid vesicles (liposomes, virosomes) or micelles. The inability to propagate hepatitis *B* virus in tissue culture provided an incentive for the production of one of the first subunit vaccines, Heptavax-*B*, which was obtained by purifying and formalin-inactivating the surface antigen from the plasma of chronic carriers of hepatitis *B*. Experimental subunit vaccines have also been produced from purified glycoproteins derived from the virus membranes of enveloped virus particles, such as parainfluenza-3, measles, rabies, influenza, respiratory syncytial virus (RSV), feline leukemia virus, bovine herpes virus1, and pseudorabies.

Immune stimulating complexes (ISCOMs) are very immunogenic and represent an interesting delivery system for glycoprotein subunit vaccines. These complexes are really advanced liposomes that look like small balls and are made from an adjuvant.

Synthetic Peptides

Chemically synthesized peptides are perceived as one alternative to conventional vaccines to elicit virus-neutralizing antibodies, or to "prime" the host for neutralizing antibody responses. A conventional approach to preparing antipeptide antibodies is conjugation of a peptide to a known *protein* or *synthetic polymer carrier*. Methods designed to avoid the use of carrier by polymerizing or cyclizing the peptides have also been reported, as has an approach known as the *multiple-antigen peptide system*, in which a small peptidyl core matrix is covalently bound to radially branching synthetic peptides as dendritic arms.

The peptide sequences chosen for synthesis in developing new vaccines correspond to antigenic determinants (epitopes) on the surface of virus particles that have been identified through chemical, electron-microscopic, and crystallographic analysis. Application of recombinant DNA technology has also allowed nucleotide sequences of viral genomes to be determined, from which primary amino acid sequences, secondary structure, and potential B-cell antigenic and T-cell (amphipathic) sequences may be deduced.

Carrier-free polypeptide vaccines are anticipated to be especially safe because they are chemically well defined, do not contain nucleic acids or extraneous proteins, and are unlikely to have any of the pathogens that might be present in serum or tissue extracts. Polypeptide vaccines are also extremely stable, withstanding thermal extremes and ambient conditions that might inactivate other types of vaccines under field conditions. This is important where target populations in underdeveloped countries may lack access to refrigeration equipment.

Synthetic peptide vaccines may be feasible for agents for which protection is based mainly on humoral (antibody-mediated) immunity and suitable conserved antigenic sites can be found. Synthetic peptide vaccines are more realistic for DNA than for ribonucleic acid (RNA) viruses, since DNA viruses mutate much less frequently than RNA viruses. Thus, antibodies against a single antigenic site of an RNA virus may allow too many escape mutants and, in contrast to DNA viruses, may require antibodies against different antigenic sites. Successful synthetic peptide vaccines have been produced for the simple, DNA-containing canine parvovirus and mink enteritis virus, which readily induce full protection of all vaccinated animals after a severe challenge with virulent viruses.

Biosynthetic Polypeptide Vaccines

The nucleic acid sequences encoding microbial antigens and chemically synthesized peptides can be cloned in bacterial and yeast plasmids or in viral vectors and expressed efficiently in prokaryotic and eukaryotic cells, either as genetically purified (cloned) proteins or as parts of fusion proteins. The immunogens made in this way may be secreted into the medium or may be expressed on the surface or in the cytoplasm of cultured cells. The antigens can then be purified from the culture medium or from cell lysates. The biosynthetic production of antigens is particularly useful when the microbial agent is difficult to grow in tissue culture, such as human hepatitis B virus, or dangerous, such as rabies or HIV.

Anti-Idiotype Antibody Vaccines

The binding site of an antibody possesses a highly distinctive three-dimensional structure (the idiotype) that can itself act as an antigenic determinant to stimulate the formation of antibodies. The antiidiotype antibodies, therefore, mimic the form of the surface antigen initially employed to induce the formation of a specific idiotypic antibody.

To prepare such vaccines, a monoclonal antibody is first produced specifically against a microbial antigen. This monoclonal antibody is injected into an animal to induce a second antibody specific to the monoclonal antibody. If the second antibody binds to the idiotype of the initial monoclonal antibody, then the antigen-recognition portion of the second antibody should be structurally analogous to the original microbial neutralization antigen. Injection of the second antibody as a vaccine will then induce antibodies against the original microbial agent. Experimental anti-idiotype antibody vaccines have been made against a number of diseases including hepatitis *B* virus, rabies virus, HIV, herpesvirus, poliovirus, and several bacterial and parasitic pathogens.

DNA Vaccines

DNA (gene) vaccines, or the direct injection of DNA plasmids that express antigens of interest, represent another approach to subunit vaccines. In animal models, DNA vaccines stimulate cell-mediated as well as antibody-based immune responses, unlike most conventional inactivated vaccines. The DNA vaccines persistently express genes encoded by the plasmid without integrating into chromosomal DNA. Gene expression from the direct intramuscular injection of DNA has been detected in mice, rats, rabbits, chicks, dogs, fish, cattle, and nonhuman primates, illustrating the diversity of species in which direct DNA vaccination without a delivery vehicle is possible. DNA vaccines have a number of advantageous properties. They can protect against viral infections in which the antibody response alone is not protective, or where there is pronounced antigenic diversity among the target strains. Protection against lethal challenges has been achieved through parenteral or mucosal immunizations, or by the use of gene guns that deliver tiny amounts of DNA-coated gold beads. The same or similar DNA plasmids can be used as vectors for subsequent immunizations, because no immune response is elicited to the vector. The development of new DNA vaccines may be pertinent to several diseases, including acquired immune deficiency syndrome (AIDS) and hepatitis *B*, and will undoubtedly receive increasing attention.

Adjuvants

Adjuvants are chemicals that significantly enhance speed, vigor, and persistence of strong antigens and the potency of weak antigens. Adjuvant formulations consist of adjuvants in suitable delivery vehicles, such as mineral or vegetable oil emulsions, squalene, liposomes, nonionic block polymer surfactants, and biodegradable polymer microspheres. Biodegradable microspheres can be especially useful as vehicles because of the ease of vaccine delivery by the oral route and the safety of the delivery vehicle (tissue compatibility and no secondary reactions). The most common microspheres range from a few micrometers to 200 m m in diameter.

The most common adjuvants used in immunological research are Freund's incomplete adjuvant, which consists of an aqueous antigen solution emulsified with an equal volume of paraffin oil/ mannide monooleate (Arlacel A), and Freund's complete adjuvant (muramyl dipeptide delivered in oil emulsions), which also contains killed mycobacteria. The complete adjuvant consistently stimulates high and long-lasting antibody responses, which can be attributed to the slow release of antigen from the site of injection. Neither form of adjuvant is acceptable for human use because both induce pain, abscess formation, granulomas, synovial joint lesions, and other adverse reactions. The only adjuvants authorized in the United States for human use are alum salts, which are both ineffective with some antigens and capable of producing adverse reactions, such as granulomas at the injection site. The emergence of subunit vaccines created by recombinant DNA technology and of synthetic peptide vaccines has intensified the need for safer and more effective adjuvants.

Cytokines and Immunomodulation

The induction of an immune response in the presence of an appropriate antigen initiates reciprocal interactions between antigen presenting cells, such as dendritic and Langerhans cells (and possibly macrophage) and T lymphocytes bearing receptors for processed forms of the antigens. The development and propagation of the immune response also involves diverse subtypes of *T* cells and *B* lymphocytes, natural killer and lymphokine activated killer cells, neutrophils, and eosinophils. The communications between these immune and inflammatory cells are mediated in large part by small proteins (about 20 kilodaltons) named cytokines. The cytokines include the interleukins (IL-1 to IL-18), gamma interferon, granulocyte-macrophage colony stimulating factor (GM-CSF), tumor necrosis factor, and transforming growth factor beta. The cytokines have various actions, for example, IL-1 (T- and B-cell maturation), gamma interferon [helper T cell (Th)1 upregulation, enhanced major histocompatibility complex expression], IL-2 (Th1 upregulation), IL-4 (Th2 upregulation), IL-10 (suppression of inflammatory cytokines, enhancement of B-cell proliferation), GM-CSF (co-migratory signals for dendritic cells), and IL-12 (stimulation of Th1 differentiation). Cytokines are species-specific and expensive, and there are concerns regarding their stability, toxicity, and potential autoimmunity. These issues have limited their use as vaccine components.

Immunomodulation refers to the ability of many adjuvants to modify the cytokine network and, hence, the immune response. Adjuvants can target macrophages of the Peyer's patches (lymph nodules of the intestine) to induce mucosal immune responses, or allow antigens to associate directly with Class I major histocompatibility complex (MHC) molecules on the cell surface to elicit Class I restricted CD8+ cytotoxic T-lymphocyte responses. Particulate antigens tend to target macrophages, and macrophages can be recruited to the site of antigen deposition through the action of cytokines. Freund's complete adjuvant and lipopolysaccharide-based adjuvants drive the responses of Th1 subtype CD4+ helper cells, while alum salt, cholera toxin, and pertussis adjuvants drive the responses of Th2 subtype CD4+ helper cells. The Th1 responses are associated with interferon gamma, IL-2, and IL-12 secretions and T cells involved with strong delayed-type hypersensitivity reactions. Th1 cells also secrete tumor necrosis factor, thereby enhancing B-cell responses that elicit IgG2a, the antibody subclass most efficient in binding the serum complement proteins that enhance antigen-antibody reactions. The Th2 cells secrete the IL-4, IL-5, and IL-10 that activate the production of high levels of IgG1, IgA, and IgE by B cells. Selection of the appropriate immunoregulatory adjuvant not only leads to an enhanced immune response but also determines the isotype of IgG, which other immunoglobulins are made, and how much CD4+-directed cell-mediated immunity is generated.

Recombinant Viral Vectors and Chimeric Viruses

Another option is the use of recombinant viruses for the delivery of vaccine antigens. Although foreign antigens may be expressed in these recombinant viruses and used as inactivated subunit vaccines, live-attenuated-viral-based vectors as vaccines have decided advantages. This is because a single vaccination without adjuvants can suffice for immunization by the live-virus vaccine. Replication of the viral-based vector within the host amplifies the amount of foreign antigen being expressed, thereby often increasing both cell-mediated and humoral immune responses to the foreign antigen. The processing and posttranslational modification of the foreign antigen within infected eukaryotic cells more closely resemble those occurring during natural infection with the pathogen from which the foreign antigen was derived. In addition, presentation of the antigen may be optimized.

Attenuated bacteria, such as *Salmonella*, as well as many different virus families have been used as vectors, including adenoviruses, herpesviruses, papovaviruses, retroviruses, and picornaviruses. However, the most extensively developed and exploited vector has been vaccinia virus, a member of the Poxvirus family. Vaccinia is a very large virus; about 5% of its DNA either is redundant or encodes genes that are not absolutely essential for virus replication in cultured cells. Thus, the capacity of vaccinia recombinants to accommodate foreign DNA is very large. Experimental vaccinia-based -vector vaccines include those with foreign DNA inserts derived from hepatitis B virus, influenza, malaria, Epstein-Barr virus, genital herpesvirus, respiratory syncytial virus, Lassa fever virus, human T-cell leukemia virus, and AIDS virus. One of the most interesting of the vaccinia-based viral vectors is the recombinant that expresses the rabies virus glycoprotein gene. The vaccinia/rabies recombinant was incorporated in bait as an oral vaccine and is presently being used for the immunization of wild life. Although vaccinia is a comparatively safe vaccine virus, immunosuppressed people can become systemically infected and die with generalized vaccinia infection. Therefore, it is difficult to imagine vaccinia being used in humans in view of the global AIDS epidemic. An additional problem arises from the induction of vaccinia antibodies during the initial usage of the vaccinia vector so that it becomes difficult to use the vectorial vaccine twice.

Recombinant DNA techniques have also made it possible to construct poliovirus type 1/type 3 antigenic hybrid viruses in cell cultures to elicit both type 1 and type 3 neutralizing antibodies. Three poliovirus serotypes are known, and each serotype displays three neutralizing antigenic sites. A type 1/type 3 chimeric virus constructed from clones of Sabin strain derivatives would be useful in the primary vaccination of children to reduce vaccine-associated cases of poliomyelitis.

Diarrheal Disease Vaccines

Diarrheal diseases are caused by the double-stranded, segmented RNA enteroviruses, by the single-stranded RNA Norwalk-type caliciviruses, and by enterotoxigenic bacteria, such as *Vibrio cholerae, Escherichia coli,* and *Shigella dysenteriae.* These pathogens cause about 2.5 million deaths each year in children under 5 years of age, especially in developing countries where food and water are frequently contaminated. Inactivated vaccines and purified toxoids have traditionally been used to prevent cholera (Table 1) and other diarrheal diseases. Recently, however, genetic engineering techniques have enabled scientists to develop new oral vaccines, such as those obtained by genetically engineering the deletion of virulence factors from *Shigella, Salmonella,* and *V. cholerae.*

Rotavirus Vaccines

Rotaviruses are the most common cause of severe diarrhea and dehyration in infants. In the developing countries, rotavirus diarrhea causes about 700,000 deaths per year. In the United States, about one-third of pediatric hospital admissions are due to rotaviruses, with about 20 deaths per year. Rotaviruses elaborate an enterotoxin (nonstructural protein NSP4) that triggers a signal transduction pathway that alters epithelial cell permeability and chloride secretion. Natural protective immunity against severe rotavirus disease is built up during the first 2-3 years of life. Studies with live attenuated oral rotavirus vaccines have also shown that the majority of severe episodes of rotavirus diarrhea are preventable by oral immunization. An oral tetravalent rotavirus vaccine was licensed in the United States in August 1998. This vaccine consists of a live-attenuated Rhesus rotavirus serotype 3 combined with human-rhesus reassortants expressing serotypes 1, 2, and 4. The federal government recommended that every Infant in the United States get the protective vaccine. Over 1 million children swallowed three doses of the vaccine, at 2, 4, and 6 months of age. However, in Octoper 1999 the recommendation for routine vaccination of United States infants was withdrawn, and the manufacturer voluntarily withdrew the vaccine from the market, after it was found that 56 children developed a painful and potentially fatal bowel obstruction within 1 week of receiving the vaccine. Several other candidate rotavirus vaccines, including bovine-human reassortants, are currently being investigated.

Oral Transgenic Plant Vaccines

These vaccines exemplify the most recent innovative approach to vaccine development. Ultimately, the vaccine designers hope to use common, uncooked, edible plants, such as bananas and tomatoes, as transgenic plant delivery systems to express the antigens of pathogenic organisms in the gut, so as to induce secretory IgA, the predominant form of immunoglobulin found in mucosal secretions. Presently, however, only model studies have been completed using transgenic potatoes and tobacco plants. Transgenic plants have many potential advantages as vaccines. They are easy to grow at minimal cost, even in the most underdeveloped countries; and unlike bacteria and animal cells, they do not require specialist media and equipment or stringent purification protocols. Transgenic plants can be conveniently self-fertilized to produce stable, true-breeding lines propagated by conventional horticultural techniques and stored or distributed as seeds. Neither sterile syringes nor nurses would be needed for vaccination.

Human and animal diarrheas were initially targeted to demonstrate the feasibility of oral transgenic plant vaccines. Transgenic potato and tobacco plants were engineered to express the heat-labile subunit of *E. coli* enterotoxin LT-B (LT stands for heat-labile toxin) and the related subunit of the V. *cholerae* enterotoxin (CT)-B, both of which have affinity for the cell surface receptor, monosialosyl ganglioside GM1. Transgenic plants expressing the capsid protein of the Norwalk calicivirus were also constructed. To optimize the plant cells' expression of the bacterial and viral proteins, a constitutive plant promoter from cauliflower mosaic virus was used to drive transcription, and a plant-specific termination sequence derived from a soybean vegetative storage protein gene was inserted into the constructs. Fecal IgA secretory antibodies and serum IgG antibodies, were induced by mice fed (via a stomach tube) transgenic plants. Mice immunized with potato LT-B were also partially protected after exposure to intact LT enterotoxin. Furthermore, in preliminary human trials, volunteers who ingested the transgenic potatoes expressing the LT-B subunit developed serum and secretory neutralizing antibodies with minimal adverse side reactions. In other studies, a protective antibody response to foot-and-mouth disease virus was induced in mice following oral or parenteral immunization with alfalfa transgenic plants expressing the foot-and-mouth disease virus structural protein VP1; and neutralizing antibodies were induced by mice immunized with leaf extracts from transgenic plants expressing the glycoprotein S polypeptides of swinetransmissible gastroenteritis coronavirus. Likewise, rabies virus glycoprotein G has been expressed in transgenic tomatoes.

Cancer Vaccines

Materials commonly termed cancer vaccines are usually immunotherapeutic preparations targeted to increase the immune response to mutated cellular proteins. They are intended to inhibit the growth of previously diagnosed cancers rather than to prevent the appearance of new, spontaneous malignant tumors. However, in cases in which infectious agents are strongly associated with cancer induction, vaccines capable of preventing cancers have been made available.

Viral hepatitis comprises common infectious diseases of humans. There are more than 350 million carriers of the hepatitis type B virus (HBV) in the world, with about 1 million in the United States. It has been estimated that HBV causes about 1 million deaths worldwide, with at least 5000 deaths in the United States, each year. Carriers of HBV are at risk of developing hepatocellular carcinomas many years after the initial HBV infection. Hepatocellular carcinoma is one of the most common cancers of the world (ranking within the top 10). In some parts of Asia, it may be the first or second most common cancer. A nationwide HBV vaccination program was implemented in Taiwan in 1984. Since the initiation of this universal vaccination program, the incidence of hepatocellular carcinoma in Taiwanese children has halved, and the mortality rate for hepatocellular carcinoma has also decreased. Vaccination programs in Korea and Gambia have also decreased the incidence of hepatocellular carcinoma.

Marek's disease virus is a cell-associated herpesvirus of poultry and causes a naturally occurring contagious malignant lymphoma in chickens. The disease has been effectively controlled for more than a decade using a live vaccine based on a non-pathogenic serologically related herpesvirus of turkeys. Cancers caused by other known *infectious agents* (such as, Epstein-Barr virus, Hepatitis C virus) may also ultimately be preventable by vaccination.

10

Biodegradation

Introduction

It is the destruction of organic compounds by microorganisms. Microorganisms, particularly bacteria, are responsible for the decomposition of both natural and synthetic organic compounds in nature. Mineralization results in complete conversion of a compound to its inorganic mineral constituents [for example, carbon dioxide (CO_2; (from carbon), sulphate or sulphide (from organic sulphur), nitrate or ammonium (from organic nitrogen), phosphate (from organophosphates), or chloride (from organochlorine)]. Since carbon comprises the greatest mass of organic compounds, mineralization can be considered in terms of CO_2 evolution. Radioactive carbon-14 (^{14}C) isotopes enable scientists to distinguish between mineralization arising from contaminants and soil organic matter. However, mineralization of any compound is never 100% because some of it (10-40% of the total amount degraded) is incorporated into the cell mass or products that become part of the amorphous soil organic matter, commonly referred to as humus. Thus, biodegradation comprises mineralization and conversion to innocuous products, namely biomass and humus. Primary biodegradation is more limited in scope and refers to the disappearance of the compound as a result of its biotransformation to another product.

Enrichment Culture

Compounds that are readily biodegradable are generally

utilized as growth substrates by single microorganisms. Many of the components of petroleum products (and frequent ground-water contaminants), such as benzene, toluene, ethylbenzene, and xylene, are utilized by many genera of bacteria as sole carbon sources for growth and energy. Microorganisms can be isolated by a small environmental sample that is placed into a flask containing a mineral salts medium with the desired growth substrate (for example, benzene). Thus, the medium is enriched to select those microorganisms (prototrophs) able to utilize the substrate for growth and to synthesize all of their amino adds, nucleic acids, vitamins, and other cellular constituents from inorganic nutrients. Organisms that are capable of metabolizing the compound but require organic growth factors (auxotrophs) are not selected by this procedure. The addition of trace quantities of yeast extract or other growth supplements may enable selection of some auxotrophs, but a high concentration of growth supplements should be avoided as it will enrich for yeast extract utilizers instead of benzene utilizers in the example above.

Cometabolism and Fortuitous Metabolism

The process whereby compounds not utilized for growth or energy are nevertheless transformed to other products by microorganisms is referred to as cometabolism. Chlorinated aromatic hydrocarbons, such as diphenyldichloroethane (DDT) and polychlorinated biphenyls (PCBs), are among the most persistent environmental contaminants; yet they are cometabolized by several genera of bacteria, notably *Pseudomonas, Alcaligenes, Rhodococcus, Acinetobacter, Arthrobacter,* and *Corynebacterium.* Cometabolism is caused by enzymes that have very broad substrate specificity.

Bacteria that have evolved pathways for the complete catabolism of aromatic hydrocarbons use these same enzymes to fortuitously transform many chlorinated analogs to more oxidized metabolites. However, not all enzymes in the catabolic pathway have broad specificity, so that a chlorinated product eventually accumulates. For example, bacteria that utilize biphenyl cometabolize polychlorinated biphenyls to chlorobenzoates, which are not acted upon by the enzymes that process benzoate. However, chlorobenzoates are further metabolized by other bacteria that have enzymes which act upon them. Although cometabolism of polychlorinated biphenyls in the environment is a very slow process, it can be greatly enhanced by the addition of analogous growth-promoting substrates, such as biphenyl, a process referred to as analog enrichment.

Microbial Consortia and Catabolic Genes

Recalcitrant compounds are not degraded by single microorganisms but by a consortium of different organisms. In some cases, the interactions may be loosely associated or asymbiotic (protocooperation) whereby each organism carries out a different step in the catabolic pathway, yet is not dependent upon the other. In other cases, the association may be symbiotic where both benefit (mutualism) or only one benefits (commensalism). Biodegradation of polychlorinated biphenyls in soil is an example of commensalism in which the organism carrying out the initial oxidation receives no benefit while another organism benefits from utilizing the products of cometabolism (chlorobenzoates) for growth. Mutualism normally involves cross-feeding (syntrophism) in which all members are dependent upon each other. It was discovered in the mid-1970s that dichlorodiphenyltrichloroethane was dehalogenated anaerobically and then mineralized aerobically by different bacteria. It has been shown that polychlorintated biphenyls that are inert to aerobic oxidations (particularly those containing six or more chlorine atoms per molecule) are dehalogenated to lesser chlorinated congeners by anaerobic bacteria in sediments. Thus, the potential exists for complete biodegradation of chlorinated aromatic hydrocarbons by a sequence of anaerobic/aerobic catabolic reactions.

An alternative strategy to mineralization by microbial consortia is to combine the genes of parental strains having the complementary parts of the catabolic pathway, into a single organism by recombinant deoxyribonucleic acid (DNA) technology or by natural enhanced mating of parental strains: Catabolic genes that are encoded on plasmids, which are small extrachromosomal elements, are much more readily transferred and expressed in other bacteria than genes coded on the chromosome. Genetic exchange of catabolic plasmids among indigenous bacteria occurs naturally in the environment.

Bioremediation

The use of microorganisms to remediate the environment of contaminants is referred to as bioremediation. This process is most successful in contained systems such as surface soil or ground water where nutrients, mainly inorganic nitrogen and phosphorus, are added to enhance growth of microorganisms and thereby increase the rate of biodegradation. The process has little, if any, applicability to a large open system such as a bay or lake because the nutrient level (that is, the microbial density) is too low to effect substantive biodegradation and the system's size and distribution preclude addition of nutrients. Bioremediation of oil spills on beaches and of ground water containing xylene, from leaking storage tanks can be enhanced by the addition of Nitrogen (N) and Phosphorus (P), which are essential macroelements of all cells.

Remediation of petroleum products from ground waters is harder to achieve than surface soil because of the greater difficulty in distributing the nutrients throughout the zone of contamination, and because of Oxygen (O_2) limitations. The latter problem is caused by the low solubility of Oxygen and by the high biological oxygen demand that occurs when the organisms are present in high numbers and actively mineralizing the contaminant. Hydrogen peroxide (H_2O_2), which is completely miscible in water and rapidly decomposes to oxygen, has been added with nutrients to overcome this problem. Another approach involves a venting system whereby a stream of air is continuously introduced at injection wells at the site. This process also brings about volatilization of the contaminants, which may be metabolized by the indigenous microorganisms as they move upward through the soil. The removal of volatile organic compounds by air stripping and enhancement of biodegradation is referred to as bioventing. This process has proved to be successful in removing chlorinated organic solvents such as trichloroethylene that are major ground-water contaminants.

Fungal Enzymes

Bacteria are the primary biocatalysts of organic contaminants, and no eukaryotic organism has ever been isolated that can utilize an aromatic hydrocarbon for growth. Nevertheless, fungi have superior ability to degrade lignin, the primary and most recalcitrant constituent of wood fiber. The extracellular enzymes, in the presence of oxygen, create highly oxidative radicals which break many of the carbon-carbon bonds and the aromatic ether bonds of lignin. The ligninases of *Phanerochaete chrysosporium* fortuitously oxidize many aromatic compounds, including diphenyldichloroethane, polychlorinated biphenyls, polyaromatic hydrocarbons, and trinitrotoluene.

Because this mode of biodegradation is nonspecific and does not proceed in a sequential manner as with bacterial consortia, it is less predictable, less controllable, and less likely to proceed to completion. It is also far less successful in soil than in culture because soil organic matter and other soil constituents also react with the enzyme-catalyzed radicals. Nevertheless, this obstacle has been partially overcome in land-farming operations by diluting soil with massive amounts of plant residues upon which the fungus is maintained or by adding the contaminated soil to reactors containing the fungus.

11

Lactic Acid

Introduction

Lactic acid is a *chemical compound* that plays a role in several *biochemical* processes. It is the *acid* that gives old *milk* its sour taste, and it accumulates in skeletal *muscles* after extensive exercise, causing muscle pain.

It is a *carboxylic acid* and an *alcohol*; its *chemical formula* is $C_3H_6O_3$ and its structure is reflected in its systematic name 2-*Hydroxypropanoic* acid.

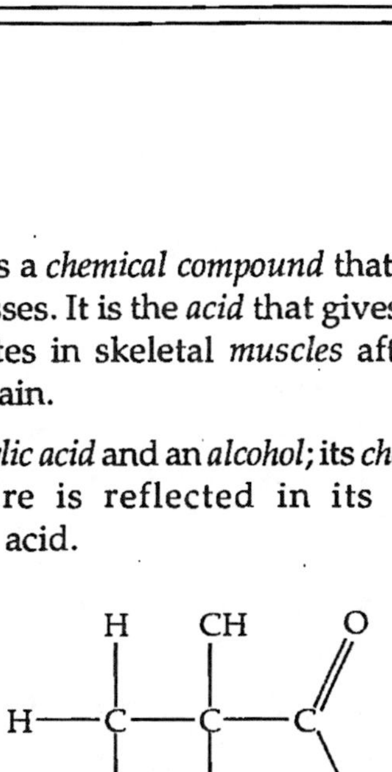

In solution, it can lose a *proton* from the COOH *carboxy group*, turning into the lactate *ion* $CH_3CHOHCOO^-$.

There are two *optical isomers* of lactic acid (and of lactate) since the central carbon atom is bound to four different groups. The first isomer is known as L(+)-lactic acid or (S)-lactic acid and the second is D(-)-lactic acid or (R)-lactic acid.

During one form of *anerobic glycolysis* or *fermentation*. L-lactate is produced from *pyruvate* via the *enzyme lactate dehydrogenase*. This conversion also oxydizes one molecude of *NADH* to NAD^+, and this is the reason for the conversion: NAD^+ has to be regenerated so that glycolysis can continue.

This lactic acid fermentation occurs in *red blood cells,* and in skeletal *muscle* during intense exertion when sufficient amounts of *oxygen* cannot be supplied fast enough. The typical lactate concentration in the blood is 1-2 *mM/L*. The *liver* takes up about 60% of the lactate and reoxydizes it to pyruvate, which is then reconverted to *glucose* in a process known as *gluconeogenesis*. The glucose enters the bloodstream and can be used by the tissues. This glucose • lactate • glucose cycle, originally described by Carl and Gerti Cori, is known as the *Cori cycle*. About 40% of the lactate is taken up by well oxygenated muscle cells and oxydized to pyruvate, which is then directly used to fuel the *citric acid cycle*.

This lactic acid fermentation is performed by *lactic acid bacteria* which are responsible for the sour taste of old milk and which are used in the production of *dairy products* such as *cheese, yoghurt* and *kefir*. Lactic acid fermentation also gives the sour taste to fermented vegetables such as traditionally cultured *sauerkraut* and *pickles* and many fermented starches such as *poi*. Lactic acid bacteria also operates in the mouth, and the resulting lactic acid is responsible for the *tooth* decay known as *caries*.

Lactic acid is also the result of *malolactic fermentation*, a process used in *winemaking* to convert sharp-tasting *malic acid* into the gentler lactic acid.

The *CAS number* of lactic acid is 50-21-5. To distinguish between the optical isomers, one can use CAS 79-33-4 for L-lactic acid and CAS 10326-41-7 for D-lactic acid and CAS 598-82-3 for an optically inactive mixture of the two.

12

Yeast and Fermentation

Introduction

Yeast are a group of unicellular *fungi* a few species of which are commonly used to *leaven bread* and *ferment alcoholic beverages*. Most yeasts belong to the *division Ascomycota*. A few yeasts, such as *Candida albicans* can cause *infection* in humans. More than one-thousand species of yeasts have been described. The most commonly used yeast is *Saccharomyces cerevisiae*, which was domesticated for wine, bread and beer production thousands of years ago.

Yeast physiology can be either obligately aerobic or facultatively fermentative. There is no known obligately anaerobic yeast. In the absence of oxygen, fermentative yeasts produce their energy by converting *sugars* into *carbon dioxide* and *ethanol* (alcohol). In brewing, the ethanol is used, while in baking the carbon dioxide raises the bread and the ethanol evaporates.

An example with glucose as the substrate is

$C_6H_{12}O_6$ (glucose) • $2C_2H_5OH + 2CO_2$

Yeasts can *reproduce asexually* through *budding* or sexually through the formation of ascospores. During asexual reproduction a new bud grows out of the parent yeast when the condition is right, then after the bud reaches an adult size, it separates from the parent yeast. Under low nutrient conditions, yeasts that are capapable of sexual reproduction will form ascospores. Yeasts that are not capable of going through the full sexual cycle are classified in the genus *Candida*.

Yeasts for leavening bread may be produced commercially or caught from the environment. Many yeasts can be isolated from sugar-rich environmental samples. Some good examples include fruits and berries (such as grapes, apples or peaches), exudates from plants (such as plant saps or cacti). Some yeasts are found in association with insects.

The use of *potatoes*, water from potato boiling, *eggs*, or *sugar* in a bread dough accelerates the growth of yeasts. Salt and fats such as *butter* slow yeast growth down. A common medium used for the cultivation of yeasts is called *potato dextrose agar* (PDA) or potato dextrose broth. Potato extract is made by autoclaving cut-up potatoes with water for 5 to 10 minutes and then decanting off the broth. Dextrose (glucose) is then added (10 g/L), and the medium is sterilized by autoclaving.

Yeast fermentations comprise the oldest and largest application of microbial technology. They are used for beer and wine fermentations and bread production. Beer brewers classify yeasts as top-fermenting and bottom-fermenting. This distinction was introduced by the *Dane Emil Christian Hansen*.

Top-fermenting yeasts (so-called because they float to the top of the beer) can produce higher alcohol concentrations and prefer higher temperatures. An example is *Saccharomyces cerevisiae*, known to brewers as *ale* yeast. They produce fruitier, sweeter, *real ale* type beers. Bottom-fermenting yeasts ferment more sugars leaving a crisper taste and work well at low temperatures. Examples include *Saccharomyces uvarum* and *Saccharomyces carlsbergensis*. They are used in producing *lager*-type beers. Brewers of wheat beers often use varieties of *Torulaspora delbrueckii*.

Winemakers use a variety of different yeasts depending on the type of wine and the condition of the *grapes*. Too high a sugar or alcohol concentration slows the growth of yeast, so for very ripe grapes with lots of sugar he or she would use a yeast tolerant of those conditions. If the yeast dies before all the fermentable sugar has been converted to alcohol, the result is a "stuck" fermentation. Some yeast is chosen because it tends to develop certain aromas, such as the distinctive "banana" smells of *Beaujolais* from *Georges*

DeBeouf. Wild yeast are naturally present on the skins of grapes, so grape juice will spontaneously ferment unless the wild yeast are arrested by cold temperature or sulphates. Depending on the strain of indigenous yeast, the result may be unpalatable or possibly more complex than if a single cultured strain were used. In general, natural yeasts are riskier than cultured, and tend to be used by tradition-oriented, *Old World*-style winemakers.

Saccharomyces cerevisiae is also known as *budding* or *baker's yeast*. It is used as a *model organism* by biologists studying *genetics* and *molecular-biology* (in particular the *cell cycle*) because it is easy to culture but as a *eukaryote*, it shares the complex internal cell structure of plants and animals.

Saccharomyces cerevisiae was the first eukaryotic genome that was completely sequenced. The yeast genome database [1] (*http://www.yeastgenome.org/*) is highy annotated and remains a very important tool for developing basic knowledge about the function and organization of eukaryotic cell genetics and physiology. Another important *S. cerevisiae* database is maintained by the Munich Information Center for Protein Sequences [2]

Another important experimental model is *Schizosaccharomycespombe or* fission yeast.

Fermentation

Decomposition of foodstuffs generally accompanied by the evolution or gas. The decomposition example is alcoholic fermentation, in which sugar is converted into alcohol and carbon dioxide. This conversion described by the equation, shown, was established by J. L. Gay-Lussac in 1815.

$$\underset{\text{Sugar}}{C_6H_{12}O_6} \rightarrow \underset{\text{Carbon dioxide}}{2CO_2} + \underset{\text{Alcohol}}{2C_2H_5OH}$$

Before 1800 the association of yeast or leaven with fermentation had been noted, but the nature of these agents was not understood. Experiments of C. Cagniard-Latour, of F. T. Kutzing, and of T Schwann in 1837 indicated that yeast is a living organism and is the cause of fermentation. This view was opposed by such leading chemists as J. von Liebig and F. Wohler, who sought a chemical rather than a biological explanation of the process. The biological concept became generally accepted following the work of Louis Pasteur, who concluded that fermentation is a physiological counterpart of oxidation, and permits organisms to live and grow in the absence of air (anaerobically). This linked fermentation and putrefaction as comparable processes; both represent decompositions of organic matter brought about by microorganisms in the absence of air. The difference is determined by the nature of the decomposable material; sugary substances generally yield products with pleasant odor and taste (fermentation), whereas proteins give rise to evil-smelling products (putrefaction).

Pasteur also discovered the lactic acid and butyric acid fermentations, and from his experiments concluded that each kind of fermentation was caused by a specific microbe. Later work supported this idea to a large extent, and considerably increased the number of specific fermentations.

During fermentation organic matter is decomposed in the absence of air (oxygen); hence, there is always an accumulation of reduction products, or incomplete oxidation products. Some of these products (for example, alcohol and lactic acid) are of importance to society, and fermentation has therefore been used for their manufacture on an industrial scale. There are also many microbiological processes that go on in the presence of air while yielding incomplete oxidation products. Good examples are the formation of acetic acid (vinegar) from alcohol by vinegar bacteria, and of citric acid from sugar by certain molds (for example, *Aspergillus niger)*. These microbial processes, too, have gained industrial importance, and are often referred to as fermentations, eventhough they do not conform to Pasteur's concept of fermentation as a decomposition in the absence of air.

13

Vinegar

Introduction

It is a food condiment containing mainly acetic acid that is produced by the bacterial oxidative fermentation of various ethanolic solutions. Vinegar, like wine and certain milk products, is one of the oldest fermented foods used by humans. Babylonian records indicate that vinegars prepared from wines and beer were widely used as early as 5000 B.C.

Types

The names for the different vinegars are based on the substrates from which they are made. These include the juices from different fruits, starchy vegetables, cereals, and distilled ethanol (ethyl alcohol). In the United States, white distilled vinegar makes up over 80% of the annual production.

Much of the white distilled vinegar is made from denatured synthetic ethanol. Nitrogen compounds, minerals, and other nutrients must be added to the alcohol medium to support the growth and metabolism of the acetic acid bacteria. Acetic acid or ethyl acetate are commonly used as the ethanol denaturants.

Most other vinegars are made from ethanolic solutions generated by a yeast fermentation. Fruit juices usually contain sufficient concentrations of hexose sugars and other yeast nutrients to support yeast growth and fermentation. Cider vinegar is produced from yeast-fermented apple juice, wine vinegar from fermented grape juice. Other juices such as pineapple, orange, and pear, as well as sugarcane syrup and molasses, can serve as fermentation substrates.

Vinegar is also produced from starchy vegetables such as potatoes and from cereals such as barley, corn, and rice. Yeasts do not ferment starch to ethanol, and therefore the starch first must be broken down (saccharified) to simple sugars, mainly glucose. Malt vinegar is made from an infusion of barley malt and other cereals in which the enzymes in the malt have converted the starch to fermentable sugars.

Balsamic vinegar from Modena, Italy, is a costly product that is highly prized by gourmets. The juice of Trebbiano grapes is concentrated about 30% by simmering, then fermented by *Zygosaccharomyces* yeasts that tolerate high concentrations of sugar. A portion of the ferment is transferred annually to a succession of wooden casks, where the acetic acid fermentation and other enzymatic changes occur. The battery of casks may be 200 years or older; thus, the harvested vinegar is of great age. About 2 liters (2 qt) of vinegar are produced from 100 kg (220 lb) of the concentrated grape must.

Vinegar sold in the United States must contain at least 4 g of acetic acid per 100 ml of solution (40-grain vinegar). Other federal specifications describe permitted color, odor, presence of trace metals, and alcohol content. Nondistilled vinegars possess distinctive colors and flavors that reflect the properties of the original substrate.

Acetic Acid Bacteria

Bacteria belonging to the genus *Acetobacter* are primarily responsible for the vinegar fermentation. Four species are recognized; they are distinguished on the basis of nutrient requirements, production of brown pigments, and tolerance to ethanol. All acetic acid bacteria require oxygen for growth and metabolism.

Gluconobacter is a closely related genus that also oxidizes ethanol to acetic acid. The genus is differentiated from *Acetobacter* by the fact that it cannot oxidize acetic acid to carbon dioxide and water. *Acetobacter* species carry out the latter oxidation, often after ethanol has been depleted.

Production Methods

A variety of methods are used for the production of vinegar. The simplest fermentations are slow and inefficient but are relatively foolproof and utilize inexpensive equipment. Submerged fermenters produce 120-150-grain vinegar in minimal time, but they are expensive and system failures can occur.

Orleans Process

This is one of the older commercial methods. The fermentation is carried out in a large cask in which holes have been drilled to permit the introduction of air. Screens are used to prevent the entrance of fruit flies. The cask has a spigot for the withdrawal of finished vinegar, and the bung (stopper) contains a tube so that fresh wine or other substrate can be added without disturbing the film of vinegar bacteria. The fermentation is initiated by mixing wine with about 20% fresh vinegar. Although the process is slow and the yields of acetic acid may be under 80%, vinegar of high quality can be produced.

Trickling, Quick Process

In this procedure, the alcoholic substrate is sprayed on the surface of a chamber packed with beechwood shavings, corn cobs, or other carrier materials that support a slime composed of acetic acid bacteria. Air is sparged through the system. Because considerable heat is generated, cooling of the chamber or recirculated vinegar stock is usually required. The vinegar stock is recirculated two or three times until the desired concentration of acetic acid is attained. Generators of this type may be operated continuously for several years or more.

Submerged Fermentation

In submerged fermentation, air is sucked into the bottom of a stainless steel tank and then broken into very small bubbles by the action of a high-speed rotor. Advantages of this system are that vinegar is produced at a high rate, about 10-fold faster than the trickling filter, and the efficiency may be as high as 94%. One disadvantage is that loss of aeration for even 1 min may result in death of the culture. Because of this problem, many vinegar producers have auxiliary electric generators as backups in the event of power failure.

Final processing steps are filtration to clarify the vinegar and pasteurization at 140-158°F (60-70°C) to arrest further bacterial growth and enzyme activity. When concentrated vinegar is desired, for example, 200-grain, this can be accomplished by freezing out some of the water from a 120-grain product.

14

Bioleaching

Introduction

Bioleaching is the extraction of specific metals from their *ores* through the use of *bacteria*.

Bioleaching is a new technique used by the mining industry to extract *minerals* such as *gold* and *copper* from their *ores*. Traditional extractions involve many expensive steps such as *roasting* and *smelting*, which requires sufficient concentrations of elements in ores. Low concentrations are not a problem for *bacteria* because they simply ignore the waste which surrounds the metals, attaining extraction yields of over 90% in some cases. These *microorganisms* actually gain energy by breaking down minerals into their constituent elements. The company simply collects the ions out of solution after the bacteria have finished.

Some advantages associated with bioleaching are:

- *economical*: bioleaching is generally simpler and therefore cheaper to operate and maintain than traditional processes, since fewer specialists are needed to operate complex chemical plants.
- *environmental*: The process is more environmentally friendly than traditional extraction methods. For the company this can translate into profit, since the necessary limiting of *sulphur dioxide* emissions during smelting is expensive. Less landscape damage occurs, since the *bacteria* involved grow naturally, and the mine and

surrounding area can be left relatively untouched. As the bacteria breed in the conditions of the mine, they are easily cultivated and recycled.

Some disadvantages associated with bioleaching are:

- *not economical*: the bacterial leaching process is very slow compared to smelting. This brings in less profit as well as introducing a significant delay in cash flow for new plants.
- *not environmental*: Toxic chemicals are sometimes produced in the process. *Sulfuric acid* and H^+ ions formed can leak into the ground and surface water turning it acidic, causing environmental damage. Heavy ions such as iron, zinc, and arsenic leak during acid mine drainage. When the pH of this solution rises, as a result of dilution by fresh water, these ions precipitate, forming "Yellow Boy" pollution. For these reasons, setup of bioleaching must be carefully planned, since the process can lead to *biosafety* failure.

Currently it is more economical to smelt copper ore rather than to use bioleaching, since the concentration of copper in its ore is generally quite high. The profit obtained from the speed and yield of smelting justifies its cost. However, the concentration of gold in its ore is generally very low. The cheaper cost of bacterial leaching in this case outweighs the time it takes to extract the metal.

The Process

The extraction of copper from its ore involves two *bacteria*, *Thiobacillus ferro-oxidans* and *Thiobacillus thio-oxidans*. In stage 1, bacteria *catalyse* the breakdown of the mineral *arsenopyrite* (FeAsS) by oxidising the sulfur and metal (in this case arsenic ions) to higher oxidation states whilst reducing *dioxygen* by H_2 and Fe_3^+. This allows the *soluble* products to dissolve:

$$FeAsS_{(s)} \rightarrow Fe_2^+{}_{(aq)} + As_3^+{}_{(aq)} + S_6^+{}_{(aq)}$$

This process actually occurs at the *cell membrane* of the bacteria. The electrons pass into the cells and are used in biochemical processes to produce energy for the bacteria to reduce oxygen molecules to water.

In stage 2, bacteria then oxidise Fe_2^+ to Fe_3^+ (whilst reducing O_2).

$$Fe_2^+ \rightarrow Fe_3^+$$

They then oxidise the metal to a higher positive oxidation state. With the electrons gained from that, they reduce Fe_3^+ to Fe_2^+ to continue the cycle.

$$M_3^+ \rightarrow M_5^+$$

The gold is now separated from the ore and in solution.

The process for *copper* is very similar. The mineral *chalcopyrite* ($CuFeS_2$) follows the two stages of being dissolved and then further oxidised, with $copper^{2+}$ ions being left.

Extraction from Mixture

Copper (Cu^{2+}) ions are removed from the solution by *ligand* exchange solvent extraction which leaves other ions in the solution. The copper is removed by bonding to a ligand, which is a large molecule consisting of a number of smaller groups each possessing a lone pair. The ligand is dissolved in an *organic* solvent such as *kerosene* and shaken with the solution producing this reaction:

$$Cu^{2+}_{(aq)} + 2LH(organic) \rightarrow CuL_2(organic) + 2H^+_{(aq)}$$

The ligand donates electrons to the copper, producing a complex - a central metal atom (copper) bonded to 2 molecules of the ligand. Because this complex has no charge, it is no longer attracted to *polar* water molecules and dissolves in the kerosene, which is then easily separated from the solution. Because the initial reaction is *reversible*, and therefore not a *displacement reaction*, it is determined by pH. Adding concentrated acid reverses the equation, and the copper ions go back into an aqueous solution.

Then the copper is passed through an electro-winning process to increase its purity: an electric *current* is passed through the resulting solution of copper ions. Because copper ions have a 2+ charge, they are attracted to the negative cathodes and collect there.

The copper can also be concentrated and separated by displacing the copper with Fe from scrap iron:

$$Cu^{2+}_{(aq)} + Fe_{(S)} \rightarrow Cu_{(S)} + Fe^{2+}_{(aq)}$$

The electrons lost by the iron are taken up by the copper. Copper is the oxidising agent (it accepts electrons), and iron is the reducing agent (it loses electrons).

Traces of precious metals such as gold may be left in the original solution. Treating the mixture with *sodium cyanide* in the presence of free oxygen dissolves the gold. The gold is removed from the solution by adsorbing (taking it up on the surface) to charcoal,

15

Cloning

Introduction

Cloning is the process of creating an identical copy of an original. A *clone* in the *biological* sense, therefore, is a multi-cellular *organism* that is *genetically* identical to another living organism. Sometimes this can refer to "natural" clones made when an organism *reproduces asexually*, but in common parlance the clone is an identical copy by some conscious design. The word was coined by the British geneticist *J.B.S. Haldane in 1963*, and is derived from the *Greek* word for "twig", *klon*.

In *biology*, cloning is used in two contexts: cloning a *gene*, or cloning an *organism*. Cloning a gene means to extract a gene from one organism (for example by PCR) and insert it into a second organism (usually via a vector). where it can be used and studied. Cloning a gene sometimes can refer to success in identifying a gene associated with some *phenotype*. For example, when biologists say that the gene for disease X has been cloned, they mean that the gene's location and DNA sequence has been identified, although the ability to specifically copy the physical DNA is a side-effect of its identification.

Cloning an organism means to create a new organism with the same genetic information as an existing one. In a modern context, this can involve *somatic cell nuclear transfer* in which the nucleus is removed from an egg cell and replaced with a nucleus extracted from a cell of the organism to be cloned (currently, both the egg cell and its transplanted nucleus must be from the same species). As the nucleus contains (almost) all of the genetic information of a lifeform, the "host" egg cell will develop into an organism genetically identical to the nucleus "donor".

However, the term *clone* is used in horticulture to mean all descendants of a single plant, produced by *vegetative reproduction*. Many horticultural varieties of plants are in effect clones, having been derived from a single individual, multiplied by some process other than sexual reproduction. As an example, some European varieties of grapes represent clones that have been propagated for over two millennia.

Clones

The modern cloning techniques involving nuclear transfers have been successfully performed on several species: (in chronological order)

- *frogs: time?*
- *sheep*: Dolly
- *rhesus monkey*: Tetra (female, January 2000)
- *pig*: 5 Scottish PPL piglets (March 2000), *Xena* (female, August 2000)
- *cattle*: *Alpha and Beta* (males, 2001)
- *cat*: *CopyCat* "CC" (female, late 2001)
- *mice:* over a dozen as of 2002
- *mule*: *Idaho Gem* (male, May 2003) and Utah Pioneer (male, June 2003)
- *deer*: *Dewey* (2003)
- *horse*: *Prometea* (female, 2003)

However, the success rate has been very low: Dolly was born after 276 failed attempts; 70 calves have been created from 9,000 attempts and one third of them died young; Prometea took 328 attempts. With certain species such as dogs and rats no successful clones have been created at all. Many people believe that attempts to perform *human cloning* would be unethical, but some scientists have publicly announced their intention to do so. Some believe the Chinese may have already done so.

A surprising development to do with aging resulted from finds that Dolly was apparently born old; she developed *arthritis* at age six. Aging of this type is thought to be due to *telomeres,* regions at the tips of *chromosomes* which prevent genetic threads fraying every time a *cell divides*. Over time telomeres get worn down until cell-division is no longer possible - this is thought to be a cause of aging. However, when researchers cloned cows they appeared to be younger than they should be. Analysis of the cow's telomeres showed they had not only been 'reset' to birth-length, but they were actually longer - suggesting these clones would live longer life spans than normal cows (but many have died young after excessive growth). Researchers think that this could eventually be developed to reverse aging in humans.

Human Cloning

Human cloning is a subject of great controversy regarding its ethical and practical consequences. A number of groups have made claims that they are working on or have already produced human clones. None of these claims has been independently confirmed.

16

Insulin

Introduction

Insulin (Latin insula, "island") is a *polypeptide hormone* primarily playing a pivotal role in the regulation of *carbohydrate metabolism*; it also takes an active part in the metabolism of fat (triglycerides) and *proteins* - it has *anabolic* properties. Insulin is used medically in some forms of *diabetes*. Type 1 diabetics depend on exogenous insulin (typically injected) for their survival because of an essentially absolute deficiency of the hormone. The first successful treatment with insulin happened in Toronto, Canada, on *January 11, 1922*.

The exact sequence of amino acids comprising the insulin molecule, the so-called *primary structure*, was determined by British molecular biologist *Frederick Sanger*. It was the first protein whose structure was completely determined. For this he was awarded the *Noble Prize in Chemistry* in 1958. In 1967, after decades of work, *Dorothy Crowfoot Hodgkin* determined the spatial conformation of the molecule, by means of x-ray diffraction studies. She also was awarded a Nobel Prize.

1. Preproinsulin (Leader, B chain, C chain, A chain); proinsulin consists of BCA, without L
2. Spontaneous folding
3. A and B chains linked by sulphide bonds
4. Leader and C chain are cut off
5. Insulin molecule remains

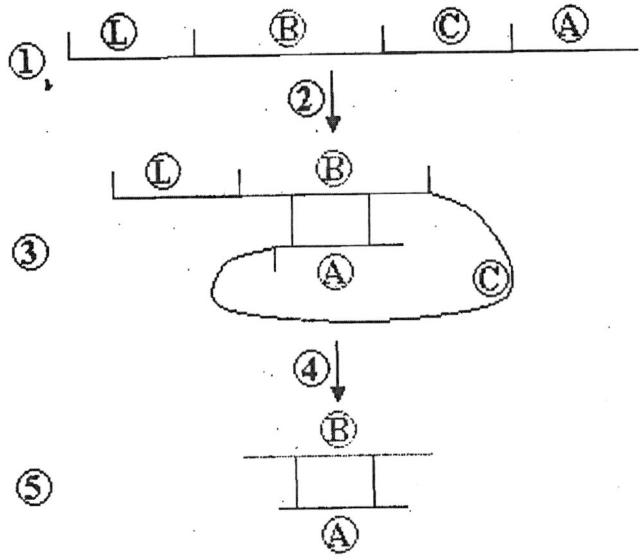

Insulin Structure and Production

Insulin is synthesized in humans and other mammals within the *beta cells* (B-cells) of the *Islets of Langerhans* in the pancreas. One to three million Islets of Langerhans (pancreatic islets) form the *endocrine* part of the *pancreas,* which is primarily an *exocrine* gland. The endocrine part accounts for only 2% of the total mass of the pancreas. Within the Islets of Langerhans, beta cells constitute 60-80% of all the cells.

Insulin is built from 51 amino acids and is one of the smallest proteins known; shorter 'proteins' are usually referred to as a *polypeptide*. Beef insulin differs from human insulin in two *amino acid* residues, and pork insulin in one residue. Fish insulin is also close enough to human insulin to be effective. In humans, insulin has a *molecular weight* of 5734. Insulin is structured as 2 polypeptide chains linked by 2 *sulfur bridges* (see figure shown above). Chain A consists of 21, and chain B of 30 amino acids. Insulin is produced as a *prohormone* molecule - *proinsulin* - that is later transformed by *proteolytic* action into the active hormone.

The remaining part is called *C-peptide*. This polypeptide is released into the blood - one C-peptide for each two insulin molecules. Since external insulins currently contain no C-peptide component, serum amounts of peptide C are good indicators of internal insulin production. C-peptide has recently been discovered to have biological activity itself; the activity is apparently confined to an effect on the muscular layer of the *arteries*.

Actions of Insulin on Cell Level and Global Metabolism Level

The actions of insulin on the global human metabolism level include:

- cellular intake of certain substances, most prominently glucose
- increase of *DNA replication* and *protein synthesis*
- modification of the activity of numerous enzymes (allosteric effect)

The actions of insulin on cells include:

- increased *glycogen* synthesis - insulin forces storage of glucose in liver (and muscle) cells in the form of glycogen; absence of insulin causes liver cells to convert glycogen to glucose and dump it into the blood
- increased *fatty acid* synthesis - insulin forces fat cells to take in glucose which is converted to fatty acids; lack of insulin causes the reverse
- increased esterification of fatty acids - forces adipose tissue to make fats (ie, triglycerides) from fatty acid esters; lack of insulin causes the reverse
- decreased *proteinolysis* - forces reduction of protein degradation; lack of insulin increases protein degradation,
- decreased *lipolysis* - forces reduction in conversion of fat cell lipid stores into blood fatty acids; lack of insulin causes the reverse
- decreased *gluconeogenesis* - decreases production of glucose from various substrates in liver; lack of insulin causes glucose production from assorted substrates in the liver and elsewhere
- increased amino acid uptake - forces cells to absorb

circulating amono acids; lack of insulin inhibits absorption

- increased potassium uptake - forces cells to absorb serum potassium; lack of insulin inhibits absorption
- aterial muscle tone - forces arterial wall muscle to relax increasing blood flow, especially in micro arteries; lack of insulin allows reduced flow

Regulatory Actions of Insulin on Blood Glucose Levels

Despite long intervals between meals or the occasional consumption of meals with a substantial carbohydrate load (eg, half a *birthday cake* or a bag of potato chips), human *blood* glucose levels normally remain within a narrow range. In most humans this varies from person to person from about 70 miligrams/deciliter to perhaps 110 mg/dl except shortly after eating when the blood glucose level rises temporarily; this is a total of about 5 grams (1/5 of an ounce) of glucose in the 5 liters (1¼ gallons) of blood in the average adult male. This *homeostatic* process is the result of many factors, but hormone regulation is the most important.

There are two groups of antagonistic hormones affecting blood glucose levels:

- hyperglycemic hormones (such as *glucagon*, *growth hormone*, and *adrenaline*), which increase blood glucose
- and one hypoglycemic hormone (insulin), which decreases blood glucose

This is because, at least in the short term, it is far less dangerous to have too much glucose in the blood than too little. Mechanisms which restore too low blood glucose (*hypoglycemia*) must be quick and effective because of serious consequences of insufficient glucose. They are mostly efficient, and symptomatic hypoglycemia is found almost entirely in diabetics on pharmacologic treatment. Such hypoglycemic episodes vary greatly in severity and swiftness of onset. In severe cases prompt medical assistance is essential, as death will result from sufficiently low blood glucose levels.

Beta cells in the Islets of Langerhans are sensitive to variations in blood glucose levels because of the presence of *glucokinase*, which responds, to glucose concentrations. If that level increases, more insulin from beta cell stores is released into the blood, and beta cell insulin production increases. When the glucose level comes down to the physiologic value, insulin release slows or stops. And, before the level of glucose drops dangerously low, hyperglycemic hormones come into play, forcing release of glucose into the blood from cellular stores.

Insulin and the Brain

Though other cells can use other fuels for a while (most prominently fatty acids), *neurons* are dependent on glucose as a source of energy in the non-starving human. They do not require insulin to absorb glucose, unlike muscle and adipose tissue and they have very little internal stores of glycogen. Thus, a sufficiently low glucose level first and most dramatically manifests itself in impaired functioning of the functioning of the *central nervous system* - dizzness, speech problems, even loss of consciousness, are common. This phenomenon is known as *hypoglycemia* or, in cases producing unconsciousness, *hypoglycemic coma* (formerly termed *insulin shock* from the most common causative agent). Because endogenous causes of insulin excess (such as an *insulinoma*) are extremely rare naturally, the overwhelming majority of hypoglycemia cases are caused by human action (eg, *iatrogenic* (meaning caused by medicine)), and are usually accidental. There have been a few cases reported of murder or attempted murder using insulin overdoses, but most insulin shock appears to be due to mismanagement of insulin (didn't eat as much as anticipated, or exercised more than expected), or a mistake (eg, 200 units of insulin instead of 20).

Misuse of any of three classes of medication are the usual causes of *iatrogenic* hypoglycemia:

- oral hypoglycemic agents (eg, any of the sulfonylureas, or similar drugs, which increase insulin release from beta cells in response to a particular blood glucose level)
- external insulin (usually injected subcutaneously)
- insulin resistance reducers (eg, one of a new class of drugs (troglitazone Rezulin was the first) which increase cellular sensitivity to insulin)

Intracellular Transformation of the Insulin Signal

There are special transport channels in *cell membranes* through which glucose from the blood can enter a cell. These channels are, indirectly, under insulin control in certain body cell types. A lack of circulating insulin will prevent glucose from entering those cells (eg, in untreated Type I diabetes). However, more commonly there is a decrease in the sensitivity of cells to insulin (eg, the reduced insulin sensitivity characteristic of Type II diabetes), resulting in decreased glucose absorption. In either case, there is 'cell starvation', weight loss, sometimes extreme. In a few cases, there is a defect in the release of insulin from the pancreas. Either way, the effect is the same: elevated blood glucose levels.

Activation of *insulin receptors* leads to internal cellular mechanisms which directly affect glucose uptake by regulating the number and operation of protein molecules in the cell membrane which transport glucose into the cell.

Two types of tissues are most strongly influenced by insulin: muscle cells (*myocytes*) and fat cells (*adipocytes*). The former are important because of their central role in movement, breathing, circulation, etc, and the latter because they accumulate excess calories against future needs. Together, they account for about 2/3 of all cells in a typical human body.

Diseases and Syndromes Caused by an Insulin Disturbance

There are several conditions in which insulin disturbance is pathologic:

- *diabetes mellitus* - general term referring to all states characterized by hyperglycemia
    - *diabetes mellitus type 1* - autoimmune-mediated destruction of insulin producing beta cells in the pancreas resulting in absolute insulin deficiency
    - *diabetes mellitus type 2* - multifactoral syndrome with combined influence of genetic susceptibility and influence of environmental factors, the best known being *obesity*, age, and physical inactivity, resulting in insulin insensitivity in cells requiring insulin for

glucose absorption. This form of diabetes is strongly inherited.

- other type of impaired glucose tolerance

- *insulinoma* or *reactive hypoglycemia*

Insulin as a Medication

Insulin is absolutely required for all animal (including human) life. The mechanism is almost identical in nematode worms (ie, *C. elegans)* and in mammals. In humans, insulin deprivation due to the removal or destruction of the pancreas leads to death in days or at most weeks. Insulin must be administered to patients in whom there is a lack of the hormone for this, or any other, reason. Clinically, this is called *diabetes mellitus type 1*.

Although it was evident to researchers in the late 1800s that there was some connection between the pancreas and diabetic symptoms, efforts to isolate the active principle (secretion or whatever) were unsuccessful. Progress was only made when it was realised that the digestive enzymes also produced by the pancreas destroyed the active material during the attempts at extraction. Many around the world came close, but the announcement of isolation of insulin from the pancreases of foetal calves (which had not yet begun the production of digestive enzymes) was made on 27 July 1921 at the *University of Toronto* (by *Frederick Banting, Charles Best, James Collip,* and *J.J.R. Macleod*. For this breakthrough discovery, Macleod and Banting were awarded the *Nobel Prize* in *Physiology* or *Medicine* in 1923. Banting and MacLeod shared some of the Prize money with the others.

Harvesting pancreases from human corpses was not possible in practice, so insulin from cows or pigs or fish pancreases was used instead. All have 'insulin activity' in humans as they are nearly identical to human insulin (2 amino acid difference for bovine insulin, 1 amino acid difference for porcine). Insulin is a protein which has been very strongly conserved across evolutionary time. Differences in suitability of beef, pork, or fish insulin preparations for particular patients have been primarily the result of preparation purity and of allergic reactions to assorted non-insulin substances remaining in those preparations. Human insulin can now manufactured, using *genetic engineering molecular biology* techniques, in sufficient quantity for widespread clinical use, much reducing impurity reaction problems. Eli Lilly marketed the first such synthetic insulin, Humulin, in *1982*. Genentech Inc developed the technique Lilly used.

There are several difficulties with the use of insulin as a clinical treatment for diabetes:

- mode of administration
- selecting the 'right' dose and timing
- selecting an appropriate insulin preparation (typically 'speed and duration of action')
- adjusting dosage and timing to fit food amounts and types
- adjusting dosage and timing to fit exercise undertaken

Diabetics give themselves insulin, usually via *subcutaneously* hypodermic injection. This is both:

- non physiologic (the pancreas releases insulin (and C-peptide) gradually into the *portal vein* instead of subcutaneously in a bolus), and
- simply a nuisance for patients to inject themselves once or several times a day

There have been several attempts to improve upon this mode of administering insulin. Many people find injection awkward and painful. There are jet injectors (also used for some vaccinations by some clinics) which have different insulin delivery peaks and durations as compared to needle injection of the same amount and type of insulin. Some diabetics find control possible with jet injectors, but not with hypodermic injection. There are also 'insulin pumps' of various types which are 'electrical injectors' attached to a semi-permanently implanted needle (a catheter). Some who cannot achieve adequate glucose control by conventional needle are able to with the appropriate pump.

Unlike many medicines, insulin cannot be taken orally. It is treated precisely as any other protein in the *gastrointestinal tract*. Like all other ingested proteins, it is reduced to its amino acid components and loses all 'insulin activity'. There are research efforts underway to develop methods of protecting insulin from the digestive tract so that it can be taken orally, but none has yet proven both safe and effective.

Inhaled insulin is under active investigation as are several other, more exotic, techniques.

An *insulin pump* is a reasonable solution for some. However there are several major limitations - cost, the potential for hypoglycemic episodes, and, thus far, no approvable means of controlling insulin delivery based on blood glucose levels. If too much insulin is delivered or the patient eats less than normal, there will be hypoglycemia. On the other hand, if too little insulin is delivered by the pump, there will be hyperglycemia. Both of these can lead to potentially life-threatening conditions. In addition, indwelling catheters pose considerable risk of infection and ulceration. Thus far, insulin pumps require considerable care and effort to use correctly. Some diabetics are able to keep their glucose in reasonable control only on a pump.

Researchers have produced a watch-like device that tests for insulin levels in the blood through the skin and administers corrective doses through pores in the skin of the patient. Both electricity and ultrasound have been found to make the skin temporarily porous. The insulin administration aspect remains experimental at this writing. The blood glucose test aspect is, at this writing, commercially available essentially as described.

Another 'improvement' would be to avoid periodic insulin administration entirely by installing a self-regulating insulin source. For instance, pancreatic, or beta cell, *transplantation*. Transplantation of an entire pancreas (as an individual *organ*) is rather difficult technically, so it is not common. Generally, it is performed in conjunction with *liver* or *kidney* transplant surgery. However, transplantation of only pancreatic beta cells is a possibility. It has been highly experimental (i.e., prone to failure) for many years, but some researchers in Alberta, Canada, have developed techniques which have produced a much higher success rate (about 90% in one group). Beta cell transplant may become practical, and common, in the near future. Several other non-transplant methods of automatic insulin delivery are being developed in the research labs as this is written. None is currently close to clinical approval.

The central problem for those requiring external insulin is picking the right dose of insulin and the right timing.

> It would be best to show this graphically. Eg, a graph of typical blood glucose levels and blood insulin levels in people without diabetes and in those with diabetes injecting themselves 1, 2, 3 or four times a day.

Physiological regulation of blood glucose, as in the non-diabetic, would be best. Increased blood glucose levels after a meal is a stimulus for prompt release of insulin from the pancreas. The increased insulin level causes glucose absorption and storage, reducing glycogen to glucose conversion, reducing blood glucose levels, and so reducing insulin release. The result is that the blood glucose level rises somewhat after eating, and within an hour or so returns to the normal 'fasting' level. Even the best diabetic treatment with human insulin, however administered, falls short of the glucose control in the non-diabetic.

Complicating matters is that the composition of the food eaten affects intestinal absorption rates. Glucose from some foods is absorbed more (or less) rapidly than the same amount of glucose in other foods. And, fats and proteins both cause delays in absorption of glucose from carbohydrate eaten at the same time. As well, exercise reduces the need for insulin even when all other factors remain the same.

It is in principle impossible to know for certain how much insulin (and which type) is needed to 'cover' a particular meal in order to achieve a reasonable blood glucose level within an hour or two after eating. Nondiabetics' bęta cells routinely and automatically manage this by continual monitoring and adjustment of insulin release. All such decisions by a diabetic must be based on general experience and training (ie, at the direction of a physician or PA) and, further, specifically based on the individual experience of the patient. It is not straightforward and should never be done by habit or routine, but with care can be done quite successfully in practice.

For example, some diabetics require more insulin after drinking *skim milk* than they do after taking an equivalent amount of fat, protein, carbohydrate, and fluid in some other form. Their particular reaction to skim milk is different than other diabetics', but the same amount of whole milk is likely to cause a still different reaction even in that same person. Whole milk contains considerable fat while skim milk has much less. It is a continual balancing act for all diabetics, especially for those taking insulin.

Medical preparations of insulin (from the major suppliers - Eli Lilly and Novo Nordisk -- or from any other) are never just 'insulin in water'. Clinical insulins are specially prepared mixtures of insulin plus other substances. These delay absorption of the insulin, adjust the pH of the solution to reduce reactions at the injection site, and so on. Some recent insulins are not even precisely insulin, but so called *insulin analogs*. The insulin molecule in an insulin analog is slightly modified so that they are

- absorbed rapidly enough to mimic real beta cell insulin (Lilly's is 'lispro', Novo Nordisk's is 'aspart'), or
- steadily absorbed after injection instead of having a 'peak' followed by a more or less rapid decline in insulin action (Aventis' version is 'Insulin glargine')
- all while retaining insulin action in the human body.

The management of the choice of insulin type and its dosage and timing should be supervised by a medical professional working with the diabetic.

Allowing blood glucose levels to rise, though not to levels which cause acute hyperglycemic symptoms, is not a sensible choice. Several large, well designed, *long term studies* have conclusively shown that diabetic complications decrease markedly, linearly, and consistently as blood glucose levels approach 'normal' patterns over long periods. In short, if a diabetic closely controls blood glucose levels (ie, on average, both over days and weeks, and avoiding too high peaks after meals) the rate of diabetic complications goes down. If glucose levels are very closely controlled, that rate can even approach 'normal'. The chronic diabetic complications incle *cerebrovascular accidents* (CVA or stroke), heart attack, blindness (from proliferative diabetic *retinopathy*), toehr vascular damage, nerve damage from diabetic neuropathy, or kidney failure from *diabetic nephropathy*. These studies have demonstrated *beyond doubt* that, if it is possible for a patient, so-called *intensive insulinotherapy* is superior to *conventional insulinotherapy*. However, *close* control of blood glucose levels (as in intensive insulinotherapy) does require care and considerable effort, for *hypoglycemia* is dangerous and can be fatal.

A good measure of long term diabetic control (approx 90 days in most people) is the serum level of *glycoslyated hemoglobin* (HbA1c). A shorter term integrated measure (over two weeks or so) is the so-called 'fructosamine' level, which is a measure of similarly glyclosylated proteins (chiefly albumin) with a shorter half life in the blood. There is a commercial meter available which measures this level in the field.

Insulin Abuse

There are reports that some patients abuse insulin by injecting larger doses that lead to mild hypoglycemic states. This is Extremely dangerous and is essentially equivalent to *suffocation experimentation*. Severe acute or prolonged hypoglycemia. can result in brain damage.

17

Organic Food

Introduction

Organic food has both a popular meaning, and, in some countries, a legal definition. In everyday conversation, it usually refers to all "naturally produced" foods, or the product of *organic farming*. As a legal term, it means certified organic. The distinction is important, as the two definitions can represent quite different products.

Types of Organic Food

Organic food, like food in general, can be grouped into two categories, *fresh* and processed, based on production methods, availability and consumer perception.

Fresh food is seasonal and highly perishable. Fresh *produce* - vegetables and fruits - is the most available type of organic food, and closely associated with *organic farming*. It is often purchased directly from the growers, at *farmers' markets*, from on-farm stands, through specialty food stores, and through *community supported agriculture* (CSA) projects.

Unprocessed animal products - organic *meat, eggs, dairy* - are less common. Prices are significantly higher than for conventional food, and availability is lower. They are still premium priced items.

To qualify as "organic", the standards for fresh food are similar to those of organic farming:

- produced without synthetic chemicals (eg: *fertilizers, pesticides, antibiotics, hormones*);
- free of *genetically modified organisms;*
- (often, but not necessarily) locally grown.

Processed food accounts for most of the items in a supermarket. Little of it is organic, and organic prices are often high, however, organic processed products are now primarily purchased from supermarkets. The majority of processed organics comes from large food conglomerates, as producing and marketing products like frozen entress and other *convenience foods* is beyond the scope of the original small organic producers.

For *processed* organic food, the general definition is:

- contains only (or a minimum percentage of) organic ingredients;
- contains no artificial *food additives;*
- processed without artificial methods, materials and conditions (eg: no chemical ripening, no *food irradiation*).

Identifying Organic Food

Definitions of organic food vary. Organics can be difficult to explain by empirical measurement and reference to "scientific fact". For one, the majority of research of the last 50 years has been focussed solely on developing and supporting chemical agriculture. And, organics is an "if it ain't broke, don't fix it" proposition, concerned in large part with what NOT to do in agriculture, rather than in devising precise formulas to identify organics. When organic food is defined by rules that include "exceptions" and "approved inputs and practices" based on scientific arguments and regulatory power, the line between organic and conventional food can become blurry.

Early organic consumers, looking for, essentially, chemical-free, fresh or minimally processed food, had to buy direct from growers: *Know your farmer, know your food* was a practical reality. Organic food at first comprised mainly fresh vegetables. Individual ideas of what exactly constituted "organic" could be developed and verified through talking to farmers and actually observing farm conditions and farming activities. Small farms can grow vegetables (and raise livestock) using *organic, farming* practices, with or without certification, and this is more or less something a direct consumer can monitor.

As consumer demand for organic foods increases, high volume sales through mass outlets, typically, supermarkets, is rapidly replacing the direct farmer connection. For supermarket consumers, food production is not easily observable. Product labelling, like "certified organic", is relied on. Government regulations and third-party inspectors are looked to for assurance.

With widespread distribution of organic food, the processed food category has also become domininant, confusing the issue further. Processed foods are more difficult to understand non-technically. Commercial preparation methods, food additives, packaging and the like are outside the direct experience of most people (including organic farmers). Traditional and minimally processed products, like *flour* and *rice, baked goods*, and *canned, frozen* and *pickled* fruits and vegetables, are easier for consumers to understand by comparison with home preparation methods, although home and mass-production techniques are quite different. For *convenience foods*, like frozen prepared dishes, cooked breakfast cereals, margarine, ingredients and methods are quite a mystery. A "certified organic" label is usually the only way for consumers to trust that a processed product is "organic".

Organic Food and Preservatives

Unfortunately, there are no natural models for preserving food the way it's found in supermarkets.

Preserving food has always been a central agricultural challenge. Today, food with a long *shelf life* is the cornerstone of the food industry, providing most of the revenue and profits. In wealthier locales, an impressive array of technologies is used to make food "last" longer: home *refrigerators* and *freezers* at the consumer end, and industrial and chemical practices applied along the food production chain, from seed to field to fridge or table.

In general, organic standards cover in detail this entire process, specifying what is an "organic" ingredient or practice. However, since there is little natural reference for preparing, for example, a precooked, frozen entree, a "certified organic" label on such an item may be hard to understand. The main ingredients are one thing, the processes and additives used to assemble and store them are quite another.

This leads to a possibility that may seem startling and impossible to apply in *developed nations*: most of what's found in supermarkets today can never be called "organic". The idea is not new, and *whole foods* 'have long been part of the *health food diet*. However, should this idea become widespread, it poses a serious threat to today's agribusiness. Therefore, there is de facto agribusiness interest in controlling the definition of "organic food", by including production practices that would facilitate food preservation.

Is Organic Food "Better"?

The bottom line consumer question is: "Is organic food significantly 'better' than regular supermarket food?" If not, less attention need be paid to understanding organic vs. conventional food. If so, consumers have to educate themselves, or risk being mislead. This area is a hotbed of controversy, and there are no conclusive answers.

The basic claims for the superiority of organic food are:

- *tastier*: Organic advocates claim organic food tastes better because of the way its produced, and because there is generally a greater variety to choose from..
- *more nutritious*: Food produced under organic conditions are somehow structurally different from chemically-raised and processed products. This pro-organic claim is beyond the scope of modern science to prove or disprove. The complex make-up of food, the effect of growing and processing methods, and the internal interaction between people and their nutrients are largely unknown. Measurements of some food components - protein, carbohydrates, fat, vitamins and minerals, and so on - only account for the most obvious factors that have been identified so far.

- *non-toxic*: Organic proponents point to potential problems with toxic residues from agricultural chemicals like pesticides. There is no argument that traces do exist, however, it is widely held that: (a) they are well in "safe" limits (as established by government regulations); (b) washing and other recommended preparation methods eliminate any risk.
- *better for the environment*: By this argument, every food purchase supports the system that delivers it: if the large-scale chemical. production methods are damaging to the environment, then people who buy these products are directly contributing to the problem.

None of these claims are widely accepted as scientific fact. There are research reports, expert opinions, and *anecdotal evidence* both supporting and refuting them. Learning more about these debates leads to clearer understanding organic food, and its potential value.

Radical Organic

Faced with inconclusive research, conflicting marketing messages, and an overall avalanche of information, some food producers and consumers who want to act now are implementing radical approaches to defining and buying organic food.

Community supported agriculture (CSA) is one such approach, that cuts out all the middlemen by having consumers partner with local farmers. CSA members prepurchase "shares" in a season's harvest, and pick up their weekly portions from distribution sites. Thus, consumers provide direct financing for farms, participate in the risks and rewards of annual growing conditions, and participate with farmers in distribution networks.

Various alternative organic standards are also emerging. They generally bypass formal certification, and provide their own definition of organic food. One such, the Authentic Food standard, proposed by leading US organic farmer Eliot Coleman, includes criteria that are incompatible with current agribusiness:

- All foods are produced by the growers who sell them.
- Fresh fruits and vegetables, milk, eggs and meat products

are produced within a 50-mile radius of their place of their final sale.

- The seed and storage crops (grains, beans, nuts, potatoes, etc.) are produced within a 300-mile radius of their final sale.
- Only traditional processed foods such as cheese, wine, bread and lactofermented products may claim, "Made with Authentic ingredients."

Particularly in developed nations, it is difficult to imagine not having the majority of products found in today's supermarkets. On the other hand, most of those products didn't exist 100 years ago, and many of them are only a few decades old.

Facts and Statistics

While organics account for 1-2% of total food sales, the organic food market is growing rapidly worldwide, both in developed and developing nations.

- World organic food sales were US$23 billion in 2002.
- The world organic market has been growing by 20% a year since the early 1990s, with future growth estimates ranging from 10-50% annually depending on the country.
- In the US:
    - "Organic products are now available in nearly 20,000 natural food stores and 73 percent of conventional grocery stores, and account for approximately 1-2 percent of total food sales in the U.S."- Feb 2003
    - Two-thirds of organic milk and cream and half of organic cheese and yogurt are sold through conventional supermarkets.
- In *Germany*: *Baby food* is almost exclusively organic, and over 30% of bread baked in *Munich* is organic.

18

Intein

Introduction

An intein is a segment of a *protein* that is able to excise itself and rejoin the remaining portions (the exteins) with a *peptide bond.*

Most reported inteins also contain an *endonuclease* domain that plays a role in intein propagation. In fact, many *genes* have unrelated intein-coding segments inserted at different positions. For these and other reasons, inteins (or more properly, the gene segments coding for inteins) are sometimes called *selfish genetic elements* but it may be more accurate to call them *parasitic*.

Inteins have also been called *'protein introns'*.

Intein-mediated protein splicing occurs after mRNA has been translated into a protein. This pre-cursor protein contains three segements - an Nextein followed by the intein followed by a C-extein.

After splicing has taken place, the result is also called an extein.

The first intein was discovered in 1987. Since then, inteins have been found in all three domains of life (eukaryotes, bacteria, and archaea).

Inteins in Biotechnology

Inteins are very efficient at protein splicing and they have accordingly found an important role in *biotechnology*.

Inteins have been engineered for particular applications such as protein synthesis, and the selective labeling of protein segments, which is useful for *NMR* studies of large proteins.

Intein Naming Conventions

The first part of an intein name is based on the *scientific name* of the *organism* in which it is found, and the second part is based on the name of the corresponding gene or extein.

For example, the intein found in *Thermoplasma acidophilum* and associated with 'Vacuolar ATPase subunit A' (VMA) is called 'Tac VMA'.

Normally, as in this example, just three letters suffice to specify the organism, but there are variations. For example, additional letters may be added to indicate a strain.

If more than one intein is encoded in the corresponding gene, the inteins are given a numerical suffix starting from 5′ to 3′ or in order of their identification. For example, "Msm dnaB-1".

The segment of the gene that encodes the intein is usually given the same name as the intein, but to avoid confusion, the name of the intein proper is usually capitalized (e.g. Pfu RIR1-1), whereas the name of the corresponding gene segment is italicized.

Split Inteins

Sometimes, the intein of the pre-cursor protein comes from two genes. In this case, the intein is said to be a split intein.

For example, in *Cyanobacteria*, DnaE, the catalytic subunit alpha of DNA polymerase 111, is encoded by two separate genes, dnaE-n and dnaE-c. The dnaE-n product consists of an N-extein sequence followed by a 123aa (amino acid) intein sequence, whereas the dnaE-c product consists of a 36-aa intein sequence followed by a C-extein sequence.

19

Genetically Modified Food

Introduction

A Genetically modified food is a food product containing some quantity of any *genetically modified organism* (GMO) as an ingredient.

Some nations have very *strong disagreement* over genetically modified organisms. For example, the European Union and Japan are willing to maintain labelling and traceability standards for GM food products, while the United States claims it violates *free trade* agreements.

The first commercially grown genetically modified food crop was a tomato created by *Calgene* called the *FlayrSavr*. Calgene submitted it to the U.S. *Food and Drug Administration* for testing in 1992; following the FDA's determination that the FlavrSavr was, in fact, a tomato, did not constitute a health hazard, and did not need to be labeled to indicate it was genetically modified, Calgene released it into the market in 1994, where it met with little public comment.

Subsequent genetically modified food crops included virus-resistant squash, a potato variant that included an organic pesticide called Bt (NB: the EPA classified the Bt potato as a pesticide, but required no labeling), strains of canola, soybean, corn and cotton engineered by Monsanto to be immmune to their popular *herbicide Roundup*, and Bt corn.

There was a brief interlude where Monsanto flirted with

introducing a technology called *terminator* into food crops, which produced plants that grew sterile seeds. Monsanto claimed this was necessary to protect their intellectual property rights, since they were *licensing* the technology to farmers, and would also have provided a measure of protection against volunteer corn carrying unwanted traits, a major concern that arose during the *Starlink* debacle.

Public outcry about the undue influence that the terminator gene would give to Monsanto, particularly in less developed nations where seed saving is more common, led to its withdrawal.

Awareness grew throughout the nineties and eventually produced a strong backlash against GM foods (discussed below), which were panned as "untested", "unlabeled" and "unsafe"; following this backlash, the *International Rice Research Institute* with funding from the *Rockefeller Foundation* developed a strain of rice enriched with *vitamin A* through genetic modification, dubbed *golden rice*. Subsequently the biotech industry touted this as a boon to poor people suffering from Vitamin A deficiency, which can cause blindness. This was condemned by GM food opponents as a ploy and a public relations move.

Many prominent environmental organizations, like *Friends of the Earth and Greenpeace*, currently consider the issue of the presence of GMOs in *conventional food* products to be a major issue - indeed Greenpeace has made it a centerpiece of their activism. In 2002, opponents placed a *measure* on the *Oregon* ballot that would have made that state the first to require labelling of GMO food.

Between *1996* and *2002*, the total surface area of *land* cultivated with GMOs has increased by a factor of thirty. Land producing GMO crops grew from 1.7 million hectares (4.2 million acres) in 1996 to 52 million hectares (128 million acres) in *2001*. The value for 2002 was 145 million acres and for 2003 was 167 million acres. *Soybean* crop represented 63% of total surface in 2001, *maize* 19%, *cotton* 13% and *canola* 5%.

Four countries represent 99% of total GM surface in 2001: United States (68%), Argentina (22%), Canada (6%) and China (3%). It is estimated that 70% of products on U.S. grocery shelves include GM products. In particular, Bt corn is widely grown, as are soybeans genetically designed to tolerate Monsanto's *Roundup* herbicide.

The US Agriculture Department estimated that 38 percent of the 79 million acres of corn planted in 2003 will be genetically engineered varieties as well as 80% of the 73.2 million acres soybeans.

Genetically Modified Food in Europe

In Europe, a series of unrelated food crises during the 1990s (e.g. the BSE (or 'mad cow' disease) outbreaks and foot and mouth disease) have created consumer apprehension about food safety in general, and eroded the public trust in government oversight of the food industry. This has further fueled widespread public concern about GMOs, in terms of environmental protection (in particular biodiversity), health and safety of consumers and the right to make an informed choice. The apprehension might also be due to the perceived novelty of GM foods, as well as cultural factors relating to food. The mishandling of the BSE crisis has left some consumers unwilling to consider "science" to be a guarantee of quality.

Although some claim genetically modified foods may even be safer than conventional products, many European consumers are nevertheless demanding that their "right to know" the content and origin of the food they consume be respected.

However, as a result of the high quantity of GMO crops, the adventitious presence of GM in imported food products (shipments of grain for food, feed and processing for example), is now thought inevitable and largely unavoidable, and usually not mentioned.

In a context of local food surplus where current GM food has little added nutritional value, the European consumer is wondering why any risk should be taken.

For these reasons, the marketing of GM food is regulated in a manner that helps to provide the necessary levels of safety, transparency and reassurance. At the beginning of the 2000's, European officials insisted that new regulations were needed to "restore consumer confidence" in the technology. These new regulations required strict labelling and traceability of all food and animal feed containing more than 0.5 percent GM ingredients. Directives, such as directive 2001/18/EC, were designed to require authorisation for the placing on the market of GMO, in accordance with the precautionary principle.

One of the features of the European system is a comprehensive *pre-market risk assessment*, a system trying to provide means for products to be followed at each stage of their production and distribution, by both transmission of accurate *information* and labelling. This *traceability* is a means to implement post-market measures such as *monitoring* and *withdrawals* (recalls). This system is not only limited to GMO products but should encompass any food product ultimately.

In GMO products, traceability is usually limited to products where transformed DNA and/or transformed protein are detectable, not to products that have been produced from GMOs but no longer appears to contain modified DNA and/orproteins. Officials stress that while traceability facilitates the implementation of safety measures, where appropriate, it cannot and should not be considered as a safety measure.

In 1999, a 4 year ban was pronounced on new genetically modified crops. At the end of 2002, *European Union* environment ministers agreed new controls on GMOs could eventually lead the 15-member bloc to reopen its markets to GM foods. European Union ministers agreed to new labelling controls for genetically modified goods which will have to carry a special harmless DNA sequence (a DNA code bar) identifying the origin of the crops, making it easier for regulators to spot contaminated crops, feed, or food, and enabling products to be withdrawn from the *food chain* should problems arise. A series of additional sequences of DNA with encrypted information about the company or what was done to the product could also be added to provide more data.

Japan and GM Food

Japan like Europe maintains labelling standards for GM food products. Japanese demand and assistance has led to a small effort to set up separate processing facility for non-GM soybeans in the U.S.

Poor Nations and GM Food

China is currently a heavy producer of GM cotton.

In March 2002, China introduced *biosafety* rules that demanded strict *labelling*, extensive documentation and government approval for food shipments. Under these new rules, all soybean shipments from the United States were briefly interrupted until interim safety certificates could be acquired.

Poor nations' agriculture officials are receiving training courses on GMO at the American Agriculture Department, with instruction in the WTO rules on GM products and benefits of biotechnology. U.S. industry groups are also providing "technical assistance" to fund initiatives that promote "science-based and' transparent biotechnology regulations" in countries such as China.

20

Biological Warfare

Biological warfaie, also known as *germ warfare*, is the use of any *organism* (*bacteria*, *virus* or other disease-causing organism) or toxin found in nature, as a weapon of *war*. It is meant to incapacitate or kill An adversary.

Biological warfare is a cause for concern because a successful attack could conceivably result in thousands, possibly even millions, of deaths and could cause severe disruptions to societies and economies. However the consensus among military analysts is that except in the context of *bioterrorism*, biological warfare is militarily of little use.

The main problem is that a biological warfare attack would take days to implement and therefore unlike a *nuclear* or *chemical* attack would not immediately stop an advancing army. As a strategic weapon, biological warfare is again militarily problematic, because it is difficult to prevent the attack from spreading to either allies or to the attacker and a biological warfare attack invites immediate massive retaliation.

History

The use of biological agents is not new, but before the 20th century, biological warfare took three main forms:

- deliberate poisoning of *food* and *water* with infectious material,
- use of microorganisms or toxins in a weapon system
- use of biologically inoculated fabrics

Biological warfare is believed to have been practiced in the Middle Ages, often by flinging victims of the *Black Death* over castle walls using *catapults*. Its use has also been documented in the *French and Indian War* when British troops distributed blankets infected by *smallpox* to *Native Americans*.

Use of such weapons was banned in international law by the *Geneva Protocol* of *1925*. The 1972 *Biological and Toxin Weapons Convention* extended the ban to almost all production, storage and transport. It is, however, believed that since the signing of the convention the number of countries capable of producing such weapons has increased.

During the *Sino-Japanese War* (*1937-1945*) and *World War II, Unit 731* of the Imperial Japanese Army conducted human experimentation on thousands, mostly *Chinese*. In military campaigns, the Japanese army used biological weapons on Chinese soldiers and civilians.

Research carried out in the *United Kingdom* during *World War II* left a Scottish Island contaminated with *anthrax* for the next 48 years.

Considerable research on the topic was performed by the *United States*, the *Soviet Union*, and probably other major nations throughout the *Cold War* era, though it is generally believed that such weapons were never used. There have been reports that *United States Army* has been developing weapons-grade *anthrax* spores at a biological and chemical weapons facility in Utah at least since *1992*. However, the United States had and maintains a stated policy of never using biological weapons under any circumstances.

Biological Weapons Characteristics

Ideal characteristics of biological weapons are low visibility, high potency, accessibility, and easy delivery.

Diseases most likely to be considered for use as biological weapons are contenders because of their lethality (if delivered efficiently), and robustness (making *aerosol* delivery feasible).

The biological agents used in biological weapons can often be manufactured quickly and easily. The primary difficulty is not the production of the biological agent but delivery in an infective form to a vulnerable target.

For example, anthrax is considered an excellent agent. We use it here because it is historically important and enough information is public that this discussion can't be a manual. First, it forms hardy spores, perfect for dispersal aerosols. Second, pneumonic (lung) infections of anthrax usually do not cause secondary infections in other people. Thus, the effect of the agent is usually confined to the target. A pneumonic anthrax infection starts with ordinary "cold" symptoms and quickly becomes lethal. Finally, friendly personnel can be protected with suitable antibiotics or vaccines.

A mass attack using anthrax would require the creation of aerosol particles of 1.5 to 5 micrometres. Too large and the aerosol would be filtered out by the respiratory system. Too small and the aerosol would be inhaled and exhaled. Also, at this size, nonconductive powders tend to clump and cling because of electrostatic charges. This hinders dispersion. So, the material must be treated with silica to insulate and discharge the charges. The aerosol must be delivered so that rain and sun does not rot it, and yet the human lung can be infected. There are other technological difficulties as well.

Diseases considered for weaponization, or known to be weaponized include *anthrax, ebola, pneumonic plague, cholera, tularemia, brucellosis, Q fever, VEE, SEB* and *smallpox*. Naturally-occurring toxins that might be used in weapons include ricin, *botulism toxin*, and *mycotoxins*.

Protection Measures

The primary civil defense against biological weaponry is to wash one's hands whenever one moves to a different building or set of people, and avoid touching door knobs, walls, the ground and one's mouth and nose. Washing literally sends the germs down the drain.

More exotic methods include decontamination, usually done with household chlorine bleach (5% solution of *sodium hypochlorite*). One useful decontamination is to leave shoes in an entranceway and make people wade and handwash in a footbath of bleach. Another useful technique is to periodically decontaminate floors and door knobs.

Medical methods of civil defense include stockpiles of antibiotics and vaccines, and training for quick, accurate diagnoses and treatment. Many weaponized diseases are unfamiliar to general practitioners.

Positive pressure shelters are possible but not costeffective except for the most impor'ant installations. This is because in most attacks, the agent will disperse in a long narrow ellipse downwind from the release point. Persons outside the ellipse will not be affected except by secondary infection. Persons within the release ellipse cannot be helped by civil defense measures. They need medical diagnosis and treatment.

Examples of Biological Warfare

Rajneeshi Salmonella Attack

In a small town in Oregon, followers of the Rajneesh Yoga attempted to control a local election by infecting a salad bar with salmonella. The attack caused about 900 people to get sick, and was thus quite effective.

2001 Anthrax Attack

In September and October of 2001, several cases of anthrax broke out in the United States in the 2001 *anthrax attacks*, caused deliberately. This was a well-publicized act of bioterrorism. It motivated efforts to define *biodefense* and *biosecurity*, where more limited definitions of biosafety had focused on unintentional or accidental impacts of agricultural and medical technologies).

See Also

- Biological Weapons Convention
- Chemical warfare
- asymmetric warfare
- biosecurity
- Weapon of mass destruction

External Links

- BiodefenseEducatian.arg
- Info on chemical and biological weapons for emergency and